THE RANGE OF THE RIVER

THE RANGE
of the RIVER

A Riverine History of Empire across China, India, and Southeast Asia

IFTEKHAR IQBAL

STANFORD UNIVERSITY PRESS
Stanford, California

Stanford University Press
Stanford, California

© 2026 by Iftekhar Iqbal. All rights reserved.

No part of this book may be reproduced or transmitted in any form or by any means, electronic or mechanical, including photocopying and recording, or in any information storage or retrieval system, without the prior written permission of Stanford University Press.

Library of Congress Cataloging-in-Publication Data
Names: Iqbal, Iftekhar author
Title: The range of the river : a riverine history of empire across China, India, and Southeast Asia / Iftekhar Iqbal.
Description: Stanford, California : Stanford University Press, [2026] | Includes bibliographical references and index.
Identifiers: LCCN 2025027449 (print) | LCCN 2025027450 (ebook) | ISBN 9781503641990 cloth | ISBN 9781503644946 paperback | ISBN 9781503644953 ebook
Subjects: LCSH: Rivers–Asia–History | River life–Asia–History | Watersheds–Asia–History | Asia–Historical geography | Asia–History–19th century | Asia–History–20th century
Classification: LCC GB1335 .I73 2026 (print) | LCC GB1335 (ebook) | DDC 950/.09693–dc23/eng/20250615
LC record available at https://lccn.loc.gov/2025027449
LC ebook record available at https://lccn.loc.gov/2025027450

Cover design: Lindy Kasler
Cover art: Unsplash
Typeset by Newgen in 10/14 Minion Pro

The authorized representative in the EU for product safety and compliance . is: Mare Nostrum Group B.V. | Mauritskade 21D | 1091 GC Amsterdam | The Netherlands | Email address: gpsr@mare-nostrum.co.uk | KVK chamber of commerce number: 96249943

For
Radiyah, Ridwan, Rizwana
My Padma-Meghna-Jamuna

CONTENTS

Figures ix

Abbreviations, Names, and Terms xi

Acknowledgments xiii

Introduction 1

1 **Brahmaputra** 20
   A Journey with the Son of God of the Universe

2 **Like an Elephant** 43
   The Many Meanings of the Irrawaddy Corridor

3 **Salween** 65
   The Angry River and the Taming of the Empire

4 **Mekong** 88
   Rendezvous with the Mother of Waters

5 **The Original River** 110
   The Red Beyond the Nation and the Empire

6 **The Long River** 127
   Yearning for the Yangzi

7 **The Organic Bridge** 152
   The Mule and the Making of Asia's Largest Transregion

Reflections 170
*"Oxen in a Flower Garden": The Postcolonial State
and the Fate of the BISMRY Commons*

Notes 183

Bibliography 223

Index 245

LIST OF FIGURES

MAP 0.1	Map of BISMRY river network.	xviii
FIGURE 0.1	A partial map of the upper BISMRY river network.	2
FIGURE 1.1	The Brahmaputra's outreaches to eastern Tibet, Burma, and Yunnan via Lohit River.	22
FIGURE 1.2	Boats at Goalunda, the terminus of the Great Eastern Railway at the confluence of the Ganges and Brahmaputra Rivers, 1895.	23
FIGURE 1.3	Abor settlements on the upper Brahmaputra basin.	32
FIGURE 1.4	A hand-drawn diagram of Brahmaputra-Yangzi trade routes connecting India, China, and Tibet.	37
FIGURE 1.5	A lone deer in the company of birds and a crocodile on a Brahmaputra tributary near Sadiya.	41
FIGURE 2.1	Map of the Irrawaddy River.	45
FIGURE 2.2	Routes from Myitkyina to Sadiya.	46
FIGURE 2.3	Upper Irrawaddy tributaries and the confluence that formed the Irrawaddy.	47
FIGURE 2.4	Steamers and native boats on the Irrawaddy, at Thayatmyo, Burma, 1907.	59
FIGURE 2.5	Everyday life of humans and elephants on the bank of the Irrawaddy.	63
FIGURE 3.1	Upper Salween–upper Mekong connections.	67
FIGURE 3.2	The Salween with outreaches from the Irrawaddy and Chao Phraya.	68
FIGURE 3.3	The Salween network.	70

x   List of Figures

FIGURE 3.4	"Hauling logs from the Salween River with the patient, powerful elephant," Moulmein, Burma, 1907.	76
FIGURE 3.5	Riverine Shan settlements in southwestern China.	77
FIGURE 3.6	"Zayat and Temple in Burma."	84
FIGURE 3.7	Monkeys and a crocodile at play as humans pass by on the Mekong [In Chantaboun].	85
FIGURE 4.1	The Simao trade network across the BISMRY valleys.	90
FIGURE 4.2	Mekong outreaches to the Chao Phraya and the Salween.	92
FIGURE 4.3	Bennet Bronson's functional model.	94
FIGURE 4.4	"Bridge over branch of many-mouthed Mekong, and native craft, Saigon, Cochin-China," 1915.	104
FIGURE 4.5	"Passing a Mule Over the Mekong."	105
FIGURE 4.6	"Launching a Boat at Luang-Prabang, on the Mekong."	106
FIGURE 4.7	"A Boat ascending a rapid on the Mekong."	107
FIGURE 4.8	Dwelling and moving by the Mekong.	108
FIGURE 5.1	The Red's connection to the Mekong-Salween networks.	114
FIGURE 5.2	A mule boarding a boat on the Red.	119
FIGURE 5.3	Water villages on the Red River in northern Vietnam in the early 1900s.	124
FIGURE 5.4	Daily life by the Red River in northern Vietnam in the early 1900s.	125
FIGURE 6.1	Route map: Shanghai to Bhamo, 1898–1899.	128
FIGURE 6.2	The Yangzi elevation between Yichang and Chongqing.	130
FIGURE 6.3	Map of a journey between Shanghai and Rangoon.	133
FIGURE 6.4	Trackers hauling a junk over the Yeh-Tan (Wild Rapid), upper Yangtze River (1905).	146
FIGURE 6.5	Entrance of the Lu-Kan Gorge, upper Yangzi.	147
FIGURE 6.6	An upper Yangzi boat, ready to sail.	148
FIGURE 6.7	Chongqing on the middle Yangzi.	149
FIGURE 7.1	Major mule breeding areas in Yunnan.	154
FIGURE 7.2	Men and mules crossing the Black River.	160
FIGURE 7.3	Men and mules checking in at a resthouse on the Mengtze-Manhao route.	162
FIGURE 7.4	Mule saddles.	164

ABBREVIATIONS, NAMES AND TERMS

Abbreviations
BISMRY: Brahmaputra-Irrawaddy-Salween-Mekong-Red-Yangzi Rivers
IOR: India Office Records, The British Library, London
Mss Eur: Private Papers (European Manuscripts), The British Library, London
TNA: The National Archives UK, Kew Gardens, London
TBL: The British Library, London

Names and Terms
Abor: The more appropriate and accepted term is "Adi"
Chao Phraya: Used interchangeably with Menam and Ping
Chaung: A stream in Burmese
Chiang Hung: Also known as Kiang Hung, Keng Hung, Sipsongpanna, Jing Hung
Chiang Saen: Kiang Hsen
Chindwin: Ningthi in Manipur
Di: The prefix "Di-" in various river names across Assam originates from the Kachari (Bodo) language, where *Di* means "river." This linguistic root is evident in names such as Diputa, Dihong, Dibong, Dibru, Dihing, Dimu, Desang, and Diku. To these may be added Dikrang, Diphu, and Digatu, all found near Sadiya, the earliest known center of Chutiya (Kachari) power and civilization.
Hui: Muslims of Yunnan, also known as the Panthay
Keng Tung: Also known as Kyaingtong, Chiang Tung, Kiang Tung

Mafus: Muleteer

Nam: In Tai languages including Thai, Lao, and Shan, *Nam* means "water" and is commonly used to denote rivers, much like "Di-" in Assam. This is evident in names such as Nam Ou (Laos) and Nam Mae Kong (Mekong River). The Thai word *Menam* literally means "mother water" and is often used for major rivers, as seen in Menam Chao Phraya. Similar linguistic patterns appear across Tai-Kadai languages.

Paak: *Paak* refers to a confluence in the Mekong basin and a muddy whirlpool in the Bengal Delta, highlighting its geographical and hydrological significance across the regions

Picul: A measurement unit equivalent to 64.47 kilometers

Wuhan: Used interchangeably for Wuchang, Hankou, and Hanyang

ACKNOWLEDGMENTS

The idea for *The Range of the River* emerged as I was completing *The Bengal Delta*—more than a decade ago—which examined the political-ecological history of the region during the colonial era. It became evident that the deltaic alluvial lands sustaining agrarian societies were elementally connected to the Tibetan-Himalayan highlands and their valleys upstream. This interplay was not unique to the Ganga-Brahmaputra-Meghna system but resonated across the Irrawaddy, Salween, Mekong, Red, Yangzi, and neighboring rivers. The interconnectedness of these systems drew me toward a broader exploration—a search for the greater range of the river. The journey was long, but wading through these rivers brought the gift of time—time to gather support, wisdom, and fellowship from many wonderful colleagues, friends, and institutions. I am deeply grateful to each of them.

Parts of the research for the book project have been shared at Tufts University, Yale University, NYU and NYU Abu Dhabi, Humboldt University Berlin, Tokyo University, Keio University, Nehru Memorial Museum and Library Delhi, Universiti Brunei Darussalam, BRAC University, North South University, IIT Guwahati, and Yunnan University. The work has also been presented at various conferences and workshops, including the ASEH Toronto; ICAS Singapore; AAS in Washington, DC; AAS-in-Asia in Kyoto; World Wide Asia conference in Leiden; and the Asian Borderlands Research Network conference in Singapore, as well as events in Kolkata, Berlin, Busan, and Ho Chi Minh City.

The feedback from colleagues and friends on these occasions has been profoundly rewarding. Among those whose intellectual engagement has been invaluable are Sugata Bose, Ayesha Jalal, Kris Manjapra, Shafiqul Islam, Ling Zhang, K. Sivaramakrishnan, Peter C. Perdue, Karl Appuhn, Harry Blair, Helen Siu, Andrew Sartori, Frederick Cooper, Mark Swislocki, David Gilmartin, Sayako Kanda, Tomoko Shiroyama, Kohei Wakimura, Firdous Azim, Sk. Tawfique Haque, Jianxiong Ma, Ambika Aiyadurai, Mandy Sadan, Li Yunxia, Anthony Medrano, Maitrii Aung-Thwin, Bin Yang, Mahesh Rangarajan, Joy Pachuau, and Nitin Sinha.

David Arnold, Dipesh Chakrabarty, Prasenjit Duara, David Ludden, John McNeill, Willem van Schendel, Anna Tsing, and Dan Smyer Yü offered extremely formative and insightful feedback on different occasions. Meetings with Anthony Reid in Singapore and James C. Scott in Toronto were extremely enlightening. As always, Jon Wilson, Arupjyoti Saikia, Rohan D'Souza, Neilesh Bose, and Sunil Amrith provided invaluable insights and the warmth of friendship. During my year-long research in Berlin, Michael Mann extended vital academic support, camaraderie, and guidance on available resources in Germany. Hermann Kreutzmann and Toni Huber generously shared their expertise and some valuable maps and references. Uwe Lübken, Vincent Houben, Hannelore Lötzke, Melitta Waligora, Elisa Bertuzzo, Sadia Bajwa, and Nadja-Christina Schneider made my intellectual and cultural journey in Germany worthwhile. I am deeply thankful to all of them.

I fondly cherish the memory of Chris Bayly in Cambridge and Dietmar Rothermund in Heidelberg, whose comments on the project during its early stages were reassuring. I am indebted to Ahmed Kamal and Sharif Uddin Ahmed for their encouragement and support during my days at Dhaka University. Thanks to Ridwan Iqbal for providing me with useful references from the University of Toronto libraries. I would like to express my gratitude to Imtiaz Ahmed, Faisal Ahmed, Tariq Omar Ali, Kazi Khaleed Ashraf, David Atwill, Rana Behal, David Biggs, Wen-Chin Chang, Joya Chatterji, Jason Cons, Gunnel Cederlöf, Banani Chakravarty, Vinita Damodaran, Ruth Gamble, Bérénice Guyot-Réchard, Naomi Hossain, Ashfaque Hossain, Faisal Husain, Annu Jalais, Niaz Ahmed Khan, Naveeda Khan, Kuntala Lahiri-Dutt, David Lewis, Kasia Paprocki, Premjith Sadasivan, Elora Shehabuddin, Esther Siyem, and Jelle Wouters for their immensely fruitful conversations and insights at various stages of this project.

At the Universiti Brunei Darussalam, I received great collegial and intellectual support, often for hours over tea and *makanan ringan*, from Minhaz Uddin Ahmed, Khairudin Aljunied, Paul Carnegie, Rommel Curaming, Kathrina Daud, Steve Druce, Ooi Keat Gin, Shafinoor Islam, Bruno Jetin,

Liam Kelley, Phan Le-Ha, Victor King, Magne Knudsen, Johannes Kurz, Asiyah Kumpoh, Rui Lopez, Shaikh Abdul Mabud, Muhammad Arafat bin Muhamad, Abdillah Noh, Md Mamunur Rashid, Sanjeev Routrey, and AKM Ahsan Ullah.

The generous one-year fellowship from the Alexander von Humboldt Foundation and the institutional support of Humboldt University Berlin provided a crucial opportunity to lay the groundwork for this project. Two subsequent Humboldt fellowships, carried out in Dhaka and Berlin, were instrumental in keeping the project on track. A timely grant from Universiti Brunei Darussalam enabled me to access the extensive resources of the British Library and the National Archives at Kew Gardens. I am deeply grateful to the Humboldt Foundation and UBD for their support. I would also like to thank the staff at the Staatsbibliothek zu Berlin, British Library, Yunnan University Library, Widener Library at Harvard, Centre of South Asian Studies Library at Cambridge, Dhaka University Library, and UBD Library for their kind assistance.

I am truly indebted to the four anonymous reviewers for their thoughtful and constructive feedback, which greatly enriched this work and made the process incredibly rewarding and enlightening as the book took its final form. On the editorial front at SUP, Marcela Maxfield guided me as I dived into the river, while Dylan Kyung-lim White helped me navigate the currents to reach the shore! Dylan's unwavering support and deeply thoughtful guidance have been instrumental in bringing this project to fruition. I must also thank Justine Nicole Sargent, Gigi Mark, Erin Ivy, and Erin Greb for their exceptional work as part of the production team, diligently contributing to the art program, copyediting, and map design.

My earliest memories of riverine stories come from my father, Khondker Sarwarul Islam. Born, raised, and employed in places dominated by the rivers of the Padma basin—where he also married my mother—he embodied how the interlacing flows of waters shaped the rhythms of life, love, work, and family. With his departure at the tail end of the COVID-19, this book becomes an ongoing conversation with the memory of rivers—of him, and with him.

Radiyah, Ridwan, and Rizwana have been my formidable critics, cool company, and unfailing allies, staying close to me through the calms and storms of the river of life. The book is dedicated to this terrific trio.

Among my nonhuman companions, I owe a great deal to Sungai Brunei (the Brunei River). Its steady flow, nourished by numerous hill streams, shaped a world alive with a thousand shades of color—of rainforests, mangroves, groomed greenery, and the double rainbows that occasionally preside over them. With its interplay of clouds, sunshine, and rain, the ever-shifting sky is brought down to earth by the unruly movement of shippers, fishers, and an array of graceful quadrupeds, tetrapods, and *Aves*. Together, they created a Bornean tableau that tied me to the far grander Tibetan riverscapes across the South China Sea. Beyond this epiphany of *Homo sapiens* dwelling and a call to embrace what remains of our planetary commons, I could hardly ask for more!

Bandar Seri Begawan
May 2025

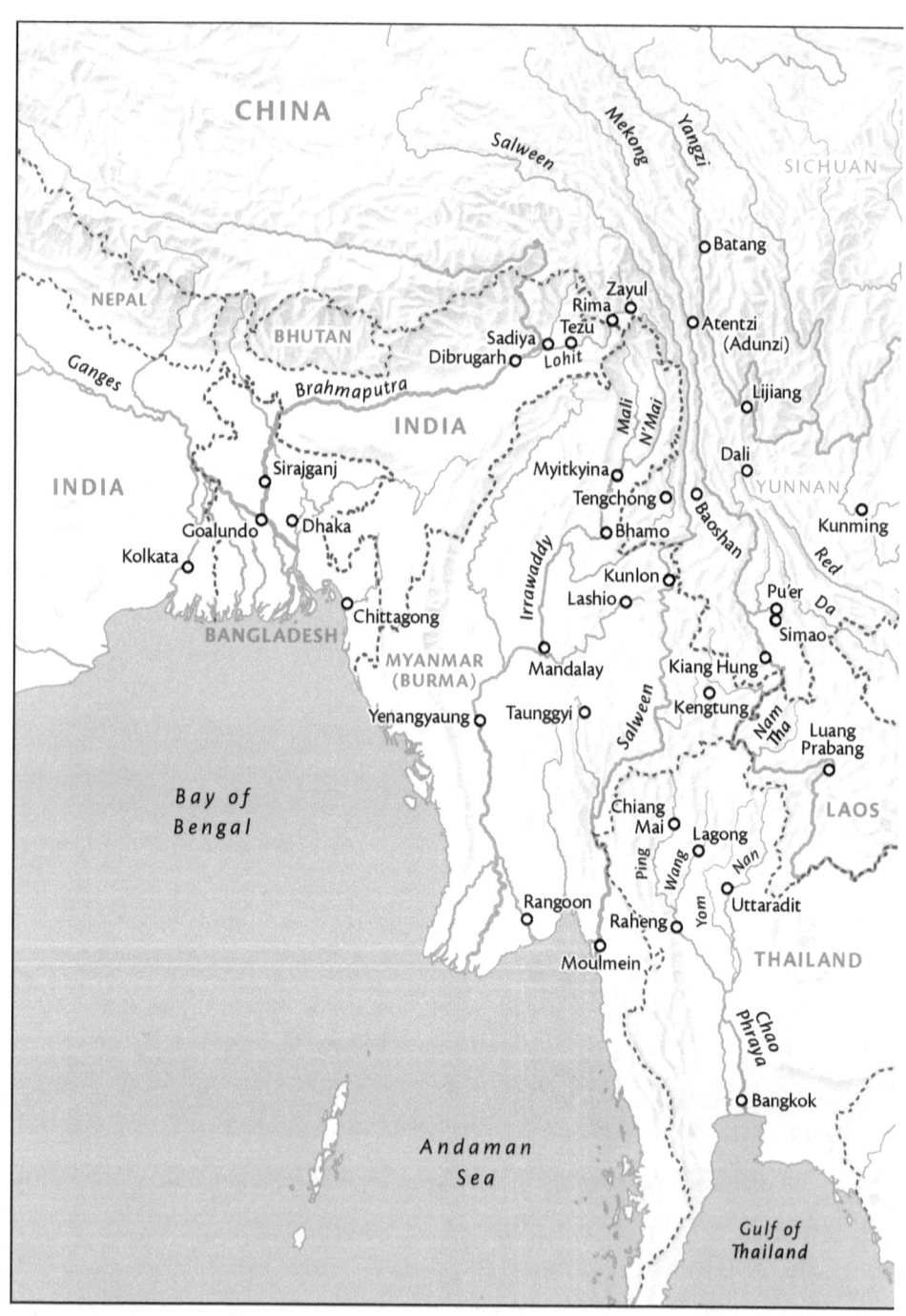

MAP 0.1: Map of BISMRY river network.

THE RANGE OF THE RIVER

INTRODUCTION

RIVERS HOLD THE KEY to a range of planetary protocols. Natural scientists identify their actions in phenomena such as the hydrological cycle, where rivers play a central role in water circulation, contributing to rainfall and groundwater replenishment and maintaining the earth's natural cooling system. They are forebearers of biodiversity, serving as vital habitats for organisms ranging from microscopic plankton to large quadrupeds, vertebrates, and invertebrates. Beyond supporting living organisms, rivers act as conveyor belts for sediments and silts, distributing nutrients that nourish soils, plants, and forests. The interconnectedness of these processes marks the river's indispensable role in sustaining the deep ecology of life on Earth.

When these planetary ecologies intersect with the human world, rivers transform into vibrant sites of mundane reality and cultural meaning. Humans have long followed rivers as natural pathways, carving settlements and trade routes along their courses. They provide essential resources for sustenance, such as water for drinking and food production. Since the mid-Holocene epoch, humans have harnessed rivers' energy to construct complex infrastructure, from dams to canals, to regulate and exploit their flows. Yet, rivers are not merely a provider of ecosystem services; they are deeply embedded in the human consciousness, serving as sources of myths, mysticism, and creative expressions. These evolving material and abstract interfaces between humanity and rivers have shaped the "riverine environment."[1]

In exploring these ranges of rivers at the intersection of planetary and human spaces, this book identifies a niche in examining six Tibetan rivers

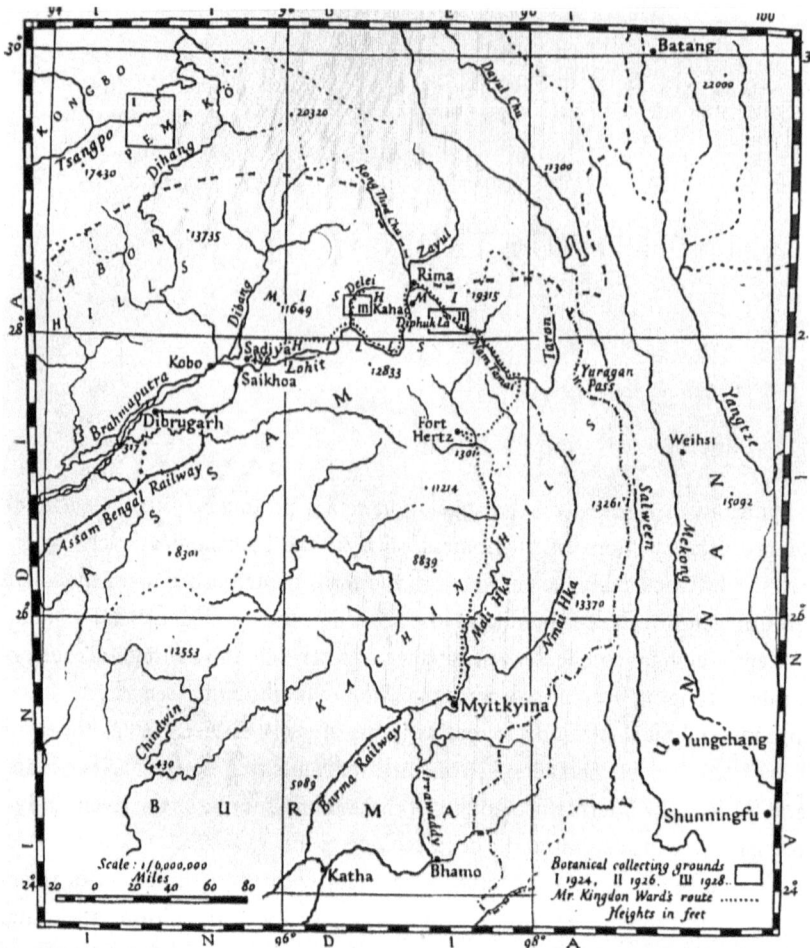

FIGURE 0.1: A partial map of the upper BISMRY river network. Source: F. Kingdon Ward, "The Seinghku and Delei Valleys, North-East Frontier of India," *The Geographical Journal*, 75, no. 5 (May, 1930): 414.

and their neighbors: the Brahmaputra, Irrawaddy, Salween, Mekong, Red, and Yangzi (Map 0.1; hereafter collectively referred to as BISMRY). These rivers are examined through three broad, interconnected phenomena. First, the spatial scale of these rivers is colossal. With nearly 4 million square kilometers of basin areas, they span eastern South Asia, mainland Southeast Asia, and southern China—constituting the largest contiguous fluvial regimes on the planet and putting a question mark on the ecologically insensitive regionalism developed during the Cold War. Second, the rivers that traversed these vast territories used and advanced a set of agencies and networks that transformed the emptiness of space into animated and interactive places. These included various confluences and tributaries of the rivers themselves, the human networks, and pack animals that operated through these networks. Finally, the rivers, through their appeal of mobility, resources, and livelihood, prompted an array of creative frictions among political forces. These waterways became contested sites where imperial ambitions, regional aspirations, and local practices intersected, often resulting in dynamic encounters over access, control, and utilization.

Against this evolving and overlapping riverine expanse, I argue that the flux of power, people, cultural practices, commodities, and the pack animals that transported them established these rivers as boundless sites of the commons. Three specific threads of a "commoning" process are examined. First, at the level of high politics, multiple imperial powers sought to establish exclusive control over one or more of these rivers. However, none achieved a monopoly, as commercial and strategic interests were constantly challenged by the physical constraints imposed by the rivers and their geomorphologically diverse landscapes. These rivers, as they shaped imperial practices along their fluvial footprints, became spaces that necessitated sharing among competing imperial forces to sustain commercial and other forms of mobility. Second, a sense of commons emerged from the contestations between imperial forces and the local ethnic groups inhabiting the BISMRY river regions. As imperial efforts to control the river networks as arteries of trade encountered the dual challenges of topographical inaccessibility and ethnic resistance, imperial powers were compelled to engage with local groups. Despite intermittent hostility, these interactions often resulted in negotiation and collaboration. Third, the BISMRY rivers fostered commoning through everyday interactions among ordinary boaters, travelers, ethnic communities, pilgrims, peddlers, merchants, smugglers, explorers, and pack animals. These exchanges often occurred directly on the rivers or in liminal spaces such as marketplaces, fairs, and sacred sites, typically located at the confluence of the rivers and their tributaries.

Before delving into the following chapters, it will be helpful to closely examine the three broad ranges of the BISMRY rivers and the contexts of the commons that emerged around them—contexts to which the late imperial order had to offer measured deference.

I
Space and Scale

The Tibetan Plateau is the launchpad for at least thirteen major river systems: Yarkand-Tarim, Amu Darya, Indus, Sutlej, Ganga, Brahmaputra, Irrawaddy, Salween, Mekong, Yangzi, and Yellow. The Red and the Pearl systems are not exactly sourced in Tibet, but neither are they separable from the connected Tibet-Yunnan-Guizhou uplands. Each of these rivers has left historic footprints in shaping politics, culture, and economy in its respective basin, inspiring several recent fine studies.[2] Among these rivers, the Brahmaputra, Irrawaddy, Salween, Mekong, Red, and Yangzi form a cluster with closer geological and hydrological links. Ferdinand von Richthofen, a German geologist, has compared the elongated ranges that extended southward from the Tibetan plateau to the fingers of the human hand through which the BISMRY flowed as a radial system.[3] This analogy syncs with the recent suggestions from paleogeologists that the Brahmaputra, Irrawaddy, Salween, Mekong, and Yangzi were once the tributaries of the Red and found their individual flows sequentially. Although this theory of giant hydrological shifts is disputed, these debates nevertheless confirm deep historical interconnections among the BISMRY rivers.[4] At around 28° 0′ 0″ N, 98° 8′ 0″ E, these rivers reached within each other's arm's length, competing for a space in and around Yunnan, and looked like the "thunderbolts in the clutch of Jove."[5]

Once the rivers left the "junction," amid mountains with a height averaging 5,000 meters, each of them moved toward widely varying destinations through a thrilling range of landscapes. The Brahmaputra moved to the West through Assam and Bengal and, after being joined by the Ganga, formed the largest delta on earth. The Irrawaddy and Salween, after crossing Chinese borders, drained almost all of Myanmar before joining the Andaman Sea. The Mekong is more transnational than any of the other BISMRY rivers, bridging China, Laos, Thailand, Cambodia, Vietnam, and marginally Myanmar on its way to the South China Sea. The Red is the shortest yet a busy route between Yunnan and the China Sea. The Yangzi, the longest of all BISMRY rivers, connected Yunnan with the East China Sea through vividly varied landscapes. Thus, the same rivers that at one point came within 200 kilometers of each other found themselves spanning 8,000 kilometers of the Indo-Pacific rim, as they emptied themselves past Dhaka, Yangon, Mawlamyine, Ho Chi Minh

City, Hanoi, and Shanghai, respectively—forming a giant watery triangle, what we would call the "fluvial Asia."

For a vast coastline receiving these rivers, which have been seamlessly entwined with the Indian and the Pacific Oceans, a *longue durée* and large-scale transregional approach is quite fitting. K. N. Chaudhuri's reimagining of the Indian Ocean space from a Braudelian lens in the 1980s was followed by the studies that zoomed in on Southeast Asia from a maritime commercial vantage point, finding masterly expression in the works by Anthony Reid. These, in turn, have paved the way for a large-scale and long-term view of Southeast Asia as a site of the "Two-Ocean Mediterranean," as Wang Gungwu suggests.[6] This flowering of Indo-Pacific studies stems from a rich tradition of oceanic studies that have largely overshadowed the role and impact of rivers. Yet, the ocean meets its limit at the shoreline, punctuated by the mouths of perennial rivers—a dynamic beautifully captured by Bengali poet Rabindranath Tagore: "Sagar bole kul mileche, ami to ar nai" (The ocean says, I have met my shore— I am no more). With historical dynamics unfolding in port cities and along the course of rivers—where currents of commerce, culture, and power converge—it is the river that shapes the contours of oceanic spaces.

On the other side of the horizon, marked by the world's largest and highest mountain ranges standing at the margins of the plains of China, India, Central Asia, and mainland Southeast Asia, Braudel looms large too. "If seas could inspire scholars to construct Braudelian regional worlds, why not the world's largest mountain ranges?" asked Willem van Schendel.[7] Braudel had already noted that parts of the Mediterranean mountains were the "refuge of liberty, democracy, and peasant 'republics'" outside of the influence of lowlanders, and communication was maintained with the latter by routes through the mountains. But Braudel ruled out such a possibility in the "impenetrable" Asian highlands, and therefore considered the people of these regions to be "autonomous."[8] The emerging field of Asian highland studies, conceptualized around the term *Zomia*, has since been greatly animated by a Braudelian parallel of "refuge of liberty" as well as by the idea of autonomy vis-à-vis the political powers ruling the plains around it.[9]

In both these spatial conceptualizations, Asia's longest rivers' terrestrial reach remains insufficiently imagined, meaning the Braudel of the Indo-Pacific is yet to meet the Braudel of the Tibetan-Himalayan-Sichuan highlands. The estimated mean annual discharges of the ten major rivers sweeping three continents into the Mediterranean is 10,000 m³/s, while the six BISMRY river discharges to the Indo-Pacific account for more than 88,000 m³/s (excluding the estimate for many other smaller perennial rivers).[10] Backed up by 40 percent of the world's freshwater reserve at the Himalaya-Tibetan Plateau, BISMRY basin areas collectively support the life of more than two billion people and countless others in the biotic community. Whereas the Mediterranean rivers are "ephemeral and seasonally intermittent," most BISMRY rivers are replenished with monsoon precipitation.[11] The Mediterranean Ocean is like a one-way magnetic field for smaller rivers, whereas BISMRY rivers roar down the Indo-Pacific but equally support a reverse magnetism toward Yunnan and Sichuan, the central gateway to China from all corners of the Indo-Pacific rim.

As historians seek to overcome the national shape of history, some well-meaning efforts have led to a reassertion of other forms of territorialization. Highlands, deltas, seas—each spatial category seems to have been prioritized at the expense of their interconnections. A critique of methodological nationalism has given way, inadvertently perhaps, to a form of methodological regionalism. Within the broader critique of the area studies as a Cold War geostrategic construct, Willem van Schendel interrogates the very idea of Southeast Asia as a region, suggesting that a territorially definitive region is untenable because of its politically informed dislocations across time and space.[12] Prasenjit Duara points to the transcending appeals of world religions to interrogate the validity of national and regional constructs.[13] Another critique emerges from the historians of trade and commerce, referring to multiple networks of people and flows of commodities that reflect the malleability of artificial constructions of regional or national space.[14] *The Range of the River* capitalizes on these recent critiques of postcolonial political and strategic territorialization and revisits the histories of transregional reaches of the BISMRY rivers.

Understanding how this extensive network of rivers spans the mountains and oceans is crucial to appreciating the spatial integrity of ecologically contiguous territories between "core" India and "core" China. As the German-American Sinologist Karl Wittfogel has argued, the arid and semi-arid regions in northern China (Yellow-Wei system) or northern India (Indus-Ganga system) were subject to complex irrigation and river-training projects under demanding imperial bureaucracies and technological interventions throughout most of their histories.[15] Despite criticism of Wittfogel's ethnocentric views, his analysis of schematic maneuvers with water resources in water-scarce societies resonates, even if marginally, with a few recent river studies.[16] If the heart of Chinese civilization beats in the Yellow-Wei system, for India this is the Indus-Ganga Valley. As this book suggests, the Brahmaputra and the Yangzi, and the rest of the BISMRY rivers in between, had a more complex journey into the southeastern Asian multiverse. These rivers were relatively less exposed to the Wittfogelian conception of exploitation, being gravitated toward transregional commercial and cultural connections across northeastern India, southeastern Tibet, mainland Southeast Asia, and southwestern China, until the cartographic constructions of the postcolonial states throttled such flows.[17] In its heyday, this massive triangle of fluvial Asia was a spatial statement against imperial China and imperial India that operated from the Yellow-Wei and Indus-Ganga Valley, respectively.

In this large-scale spatial context, did the rivers mediate the conversations between the "autonomous" mountains and "cosmopolitan" coasts? What was the nature of contestations and engagements that flowered along and across their valleys and plains? How did the fluvial Asia—which emerged out of layered assemblages of various human and nonhuman actors—carve itself a transregional commons that sits oddly with the fragmented postcolonial state geobodies? This book is an attempt to address these questions.

II
Agency and Network

"A current of fresh water, flowing in a Bed or Channel, from its source to the sea"— John Robison, a professor of natural philosophy at the University of Edinburgh, provided this simple definition of a river in 1822. But in this modest rendition, Robison saw the river as a "sensible agent of nature" as it provided access to the riparian people "of their fertile banks all their abundance."[18] Beyond the agrarian world in which Robison defined a river, the river's agency is also marked in trade and commerce that moved people and products far and wide, as Frank Kingdon Ward, an avid botanist, noted a century later: "We cannot divert trade in Asia while men live where they do live, migrate as they do migrate, while deserts and mountain ranges and rivers are where they are—it still flows along accustomed routes, and will continue to do so when the pyramids lie in the dust."[19]

Ward's remarks in the early 1920s were largely a reflection of his personal observations of eastern Tibetan rivers and pre-date the Annales School's presumption about nature's clinching power. This version of nature's "agency," as suggested by Robison and Ward, may be construed as another name for environmental determinism, but discourses on nature's agency in recent times are replete with the suggestion that there is a very thin line between agency and determinism, if we consider what in nature is moving and interacting collaterally. The Himalayas, plains, coasts, and the rivers through these landscapes speak to each other as they move and as they harbor the lives of humans, silt, fish, minerals, microbes, and all the rest. In this collateral mobility, seasonal changes in rainfall, evaporation, and tide and ebb ensure that the rivers have language, moods, and memories. "Today the river remembers what the flows were like yesterday," suggests Sean Fleming.[20] But if the language of rivers has always been audible to the poetic minds of those like Rabindranath Tagore or Ralph Waldo Emerson, historians have also begun to sense the river's "multivocality" as manifested through the march of time. If there is an agency of the river, it is found as much in its largesse as in its power to embrace other entities in nature through its ability to communicate and facilitate their mobility.[21]

The mobility and flows at the site of the river are in a large part facilitated by its network of tributaries, subtributaries, rivulets, and smaller streams. Each large river, like the Pied Piper of Hamelin, called on these smaller streams to join its main body. To the hydrologists and environmentalists, these wider river networks are key to biodiversity, demographic dynamics, and resource management, as well as ecological corridors for the movement of human and equal-to-human entities.[22] These fluvial networks, especially the confluence of each of these water bodies with the main river, often become a nodal point for mobility and interactions and a hub of societal energy. The river's performance in facilitating mobility is at its best in such confluences, which could be considered what Bruno Latour calls "actants," in the sense that it animates other actors like human or nonhuman, animate or inanimate.[23] The depth and breadth of the intermobility around the confluences are also understood from the perspective of the South Asian pantheon, in which, for example, the supreme Hindu deity of creation and transformation, Shiva, is known as the "Lord of the Meeting Rivers" (*sangama* or confluence). This has powerful symbolic and practical implications for the emerging field of environmental humanities.[24]

Each one of the BISMRY rivers was a total of many *sangama*, and the more *sangama*, the more the mobility and entanglements of people, commodities, and cultures. Some of the signature BISMRY *sangama* were Bhamo on the Irrawaddy, Sirajganj and Sadiya on the Brahmaputra, Luang Prabang on the

Mekong, Manhao on the Red, and Wuhan and Chongqing on the Yangzi (Map 0.1). While tributaries ran to their confluence with their respective main river, the beginning points of some of these tributaries happened to be close to the network of the next main neighboring river. The Brahmaputra's Lohit reached close to Irrawaddy's Mali and Chindwin, the Irrawaddy's Taping and Shweli edged up to the Salween, and so on. The translocal bonding of the water spaces thus evolved into a transregional network. Sometimes these networks were so intricate, especially in the closely flowing upper streams, that it was difficult to recognize which tributaries were meeting which river. As late as the turn of the twentieth century, debates continued as to the exact identity of some of these rivers. The Irrawaddy was considered identical with the Brahmaputra, and doubts were cast on whether Nag Chu Kha, assumed to be a feeder of the Salween, was a branch of the Mekong or the Yangzi.[25]

Yunnan stands out as the mega nodal point, around which the BISMRY rivers spread out like a necklace and cut across the territories that span South Asia, Southeast Asia, and China. Each of these rivers had a different topography, giving rise to different political, economic, agrarian, and cultural trajectories, yet spatially they were all connected to Yunnan. Each river has had a fluvial journey toward the ocean as well as toward Yunnan—as a transit point, a cosmopolitan space, and a network of circularity.

These layered river networks had obvious physiographical limits. Considerable portions of the BISMRY river flowed through high mountain ranges with limited or no navigability. But the lack of navigability does not dislodge the river as a site of mobility, because the unnavigable zone remains as attractive as a signpost and as a pathway to the next navigable point. In the case of the BISMRY basins, the problem of physical disconnects between the rivers was amply compensated by pack animals, especially mules. Here the agency of the river meets with the agency of a quadruped with special mobility skills. By the turn of the twentieth century, a rough estimate found the number of mules on the move in Yunnan to be between 50,000 and 100,000, a conservative figure that remained stable until World War II, when modern highways for motor vehicles were constructed. With mules, along with ponies and oxen, the difficulties of the high and rugged landscape between the river networks were bridged in a way that mules also did in the Mediterranean mountain zones. Braudel noted that the mule traffic contributed to the "unity of the Mediterranean region," and that "river traffic and mule trains cooperated on the journey up river."[26] Unlike those around the Mediterranean, however, the mules in the

BISMRY transregion didn't have to come all the way to the coast, because at some point each of these large rivers became navigable. But it did the most needed trick at the zones of "impenetrability" in and around Yunnan and Sichuan. Without mules, intervalley mountain passes would have been desolated and disjointed spaces; without rivers, the mules would have no purpose for their existence. Ultimately, the mule was an indispensable agency that, in collaboration with humans and rivers, bridged Asia's toughest intervalley zones. It is in this context that I refer to the mule as an "organic bridge."

III
Dialectic of Encounters

The network and agency of the river and the mule are only complete with their entanglement with the human network and mobility in the zones of human actions, where "dialectical interchanges" between humans and nature take place.[27] These interchanges were most visible in the mobility of people along the river networks in the long history of this transregion. In providing an outline of ethnic migration along the BISMRY "radial system," Kingdon Ward notes that diasporic mobility originated in a "common cradle" at the apex of a triangle formed by the headwaters of Irrawaddy, from where three routes of migration emerged: Shans and the Mon Khmers used the routes that went southeast into the Mekong and Yangzi Valley; members of the Tibetan-Burman family moved south in Burma along the Irrawaddy Valley; the Abors, Mishmis, and Nagas, among other groups, spread out southwest along the Brahmaputra. The "march to the south" along the water courses remains a standard narrative of mobility in premodern times. Between the cartographically visible large rivers that defined this vast landscape, there were numerous medium to small water bodies including rivulets, streams, tributaries, branches, and confluences that formed sites for what Toni Huber terms "micro-migration."[28] Instances of both macro- and microlevel migration are evident in the observations of a Bengali traveler in northeastern India during the 1870s. He noted that the Mishmis of eastern Assam shared ancestral ties with the Yunnanese, the Garos residing between the Surma and Brahmaputra Valleys were perceived as "dangerous" akin to the Nagas, and the Khasis to the east of the Garo Hills adopted the identity of "Bhutia Saheb" upon converting to Christianity.[29]

Mobilities southward from the Tibetan plateau resulted in the expansion of similar linguistic features. For example, what is *nawadun* for a boat in southeastern Tibet or Nepal is *naaw* or *nowka* in the Bengal Delta.³⁰ Such a multisite itinerary of cultural and linguistic lifeworld often accompanied an expanding agrarian domain, especially of domesticated rice culture, which was believed to be sourced in the region stretching from the Brahmaputra to the Yangzi Valley.³¹ The Austro-Asian Mons inhabiting the Pegu River basin and the Irrawaddy Delta trace their origin back to the Yangzi Valley in southern China from where they believe they originally migrated, the very name "Yangzi" being absorbed by the Chinese from the Mon language.³² Vertical linguistic mobility along the rivers was complemented by horizontal intervalley mobility, as seen in the similarity of the Khasi language spoken by the Khasi on the Brahmaputra system and Palaung-Wa spoken on the Mekong River, in terms of vocabulary as well as sentence formation.³³ Another form of spatial movement took a circular turn, in which, for example, Buddhist missionaries traveled along the mercantile routes that started from Bengal ports and connected China via the Irrawaddy. The word *Tsein* that denotes Burma-China frontiers may have derived from *Ceen*, by which China is known in Bengal.³⁴ Bin Yang and Radhika Seshan examine in detail the intricacy of these transregional connections in premodern times across the river and land routes that spanned Yunnan, Sichuan, Myanmar, Tonkin, Assam, and Bengal.³⁵

These deep-historical flows and interactions across and along the BISMRY rivers took a new turn in the course of the nineteenth and first half of the twentieth century, which is the timeline for this book. By the 1850s, the port cities of Calcutta, Chittagong, and Rangoon on the one hand and Canton and Shanghai on the other were in the grip of British imperial forces. This meant that for the first time in its history, this transregion was pressed by a single maritime empire from places as disparate as the East China Sea and the Bay of Bengal, leading to enhanced connectivity across northeastern India, Burma, and China.³⁶ Rivers were central to this imperial outreach.

The British imperial journey in the BISMRY transregion began in the Ganga-Brahmaputra Delta in the mid-eighteenth century. With the formal annexation of upper Assam in 1838, most of northeastern India, whose lifeline was the Brahmaputra, came under British control. This annexation coincided with British expansion in Burma, which began in the early nineteenth century as part of broader efforts to extend their influence in Southeast Asia. The First Anglo-Burmese War (1824–1826) resulted in the British annexation of Arakan (Rakhine) and Tenasserim (Tanintharyi) regions. The Second

Anglo-Burmese War (1852) led to the capture of Lower Burma, including the port city of Rangoon (Yangon). The final phase of expansion occurred during the Third Anglo-Burmese War in 1885, which culminated in the annexation of upper Burma and the incorporation of all of Burma into British India by 1886. With this annexation, the entire Irrawaddy and the Salween River basin beyond the Chinese borders came under British influence.

The British imperial expansion in South Asia and Burma largely mirrored the French advance in mainland Southeast Asia, albeit starting somewhat later. The French first took control of Saigon in 1859, signaling larger efforts to establish control in Vietnam. By 1862, the Treaty of Saigon was signed, ceding control of several provinces, including Saigon, to the French. Their control of Tonkin or northern Vietnam came with the capture of Hanoi in 1883 and the Sino-French Treaty of Tianjin in 1885, eventually leading to the formation of French Indochina, which included Tonkin, Annam, Cochinchina, Cambodia, and later Laos. These changes brought the Mekong and the Red Rivers under French influence.

The seeds of "informal" empires in southern China, with the Yangzi at its heart, began with the Treaty of Nanking following the First Opium War (1839–1842), which forced China to cede Hong Kong and several ports, including those on the Yangzi. A series of unequal treaties under the shadow of "gunboat diplomacy" turned the Yangzi into a grand site of activities by imperial powers including the British, French, German, United States, and eventually the Japanese.

As the European maritime empires—presenting themselves through both horizontal and vertical lines of connections—stoked new spatial interests, the late Qing administration and the nationalist government in China sought sway over these waterways to keep regional and sea-borne trade channels accessible to them. The resultant Chinese contestations and compromises with external forces around these river basins brought into life a large-scale but connected field of imperial interactions and flows, despite fluctuating and messy political relations.

Regardless of imperial advances from all around, however, each of the river systems was unlikely to be controlled by any single political force. Donald Worster's seminal works on the massive exploitation of rivers by modern empires have inspired a rich body of scholarship, with significant departures emerging only more recently. Corey Ross, noting how the layered entanglements of hydrosphere and imperial power gave birth to a certain "aqueous legacy," suggests that despite colonial claims of mastering water, it was never fully controlled—least of all in the "volatile hydrographic regimes of Europe's tropical colonies."[37] *The Range of the River* echoes these thoughts. In doing so, it argues that such dynamics should not be framed merely in terms of the instrumentality of imperial power or its limits. Instead, they must be understood through the diverse local political, commercial, cultural, and even psychological forces that the river itself exerted upon the empire—as if there was also an empire of rivers that took on the empire of men.

Whereas the navigable zones of the river fell to the relatively firmer grip of the empires, mountains and valleys around the fast-flowing upstream remained the preserve of numerous ethnic groups. Imperial authorities weakened to the point where the rivers ceased to be navigable for steam vehicles, but the imperial urge to connect to the next river systems never waned. In these unnavigable terrains, local ethnic groups reigned unchecked, held on to their terms, and enjoyed an upper hand in encountering the imperial forces and flows. In northern Brahmaputra, the resultant state interactions with Abor and Mishmi, among others, were a reflection of the dialogic power of the Brahmaputra and how it created a stake for a range of ethnic families. The existence of smaller but intricate land-water routes along the Irrawaddy, Salween, Chao Phraya, and Mekong meant that the Shan, Kachin, Karen, Lao, and Yao, among many other ethnic groups, had their way of securing their interests while engaging with imperial networks, as they guarded the upper parts of these river corridors with Yunnan. Similarly, the upper Red River saw "Black Flags" and many local ethnic groups entangled with external forces. On the long, torturous upper Yangzi there were regional political forces like the Taiping and the Hui, who guarded different river nodal points against Chinese imperial dominance until they were crushed by shifting alliances between the Chinese and European powers.

Yet the staunch local resistance to the maritime powers advancing to the upper parts of the BISMRY river reflected neither a terrestrial autonomy by design, nor a desire to close the doors to the wider world, as the proponents of a distinctive Zomian highland space would argue. Their resistance to imperial penetration may be read as a discursive means to facilitate commercial flow on their own terms. Rather than fleeing from the surrounding plains, these people invited the maritime imperial forces to play in their home ground—denoting a sort of rooted mobility, being connected to the ocean through their nearest river network, however far the ocean might be. As John McNeill shows, in the course of the nineteenth and the first half of the twentieth century, ethnic mobility across the Mediterranean mountains allowed people to exploit resources and embrace contacts from the outside world, leading to slow evolutionary change but "with revolutionary consequences" in terms of greater commercial exchanges.[38] In the BISMRY transregion, the choke points with the toughest mountain terrain proved to be the vital corridors of disparate flows of culture, commodities, pack animals, and humans. Instead of being the furthest backwater of the Indo-Pacific, the interior of these regions became as immensely interactive, if not as busy, as the coastal ports.

History is not just about place making, but also about holding on to it. Maritime empires sought to navigate upstream to reach China as much as the people of the upper river regions aspired to stay connected to the oceanic flows via the rivers. Rather than seeking refuge from political subjugation and economic exploitation, these local inhabitants assertively developed parallel endogenous networks that worked in tandem with the imperial network, as if to pass the next baton in a relay race. This was an undeclared protocol of encounter between the empire and the small bastions of local powers, as both needed each other and had to follow the river's footprints. Everything was in flux, and mobility within the river's network was therefore constant, prompting Kingdon Ward to suggest, "We have, in fact, long since, reached a stage in the world's history when, to the naturalist at least, every man, beast, and plant, is a foreigner."[39] There were fringes and boundaries where these river-based mobility networks met and mingled, leading to occasional contestations and conflicts, but the river remained a constant referral, a signifier, a meeting place, and a crosshair of pathos and pathways—something that Anna Tsing would call "friction."[40] This interpermeability among river-based networks is where the evolving heart of the commons beat.

IV
Contour of the Commons

The debates around the idea of the commons as a communitarian space have been largely using statist concepts like "governance," "individual," "resources," and "property rights," among others. Hardin's idea that the commons is a "tragedy," because it is everyone's property and no one's responsibility, is effectively countered by Elinor Ostrom, who shows that polycentric communal management of natural resources is possible because primary stakeholders are capable of practicing the rules of sustainable engagement with the commons.[41]

These debates on the tragedy of the commons, which tend to substantialize physical bodies as resources rather than as vibrant interactive processes, are linked with several historical dynamics that have unfolded, especially since the emergence of the postcolonial state. As the idea of state sovereignty was coupled with the securitization of national territories, natural bodies like rivers were fated to be objectified. Capitalist investment, private property laws, and economic development policies further entrenched this perspective. These undercurrents facilitated the commodification and exploitation of nature, prioritizing rents and profits over communal stewardship and sustainable practices. Consequently, the relational and communal aspects of commoning are overshadowed by a focus on control, ownership, and resource extraction—all tending to legitimize the static rather than the processual aspects of the commons. Jawaharlal Nehru, India's first long-serving prime minister, encapsulated this spatially bounded resourcification: "What India did with India's rivers was India's affair."[42] Similar was the case with the rivers in Mao-era China where, especially in the case of the Yellow River, planism and developmentalist ideologies, coupled with nationalism, decentered the river space as an intimate commons.[43] The marginalization of the river in the consciousness of the political elite in the postcolonial era offers useful clues to its decommonization—a topic that will be explored in detail in the concluding chapter of this book.

While debates on commons continue to resonate with welfare economics, more recent studies of commons seek to go beyond the critique of its foreclosure, either as an inevitable tragedy or communal homeostasis, and open it up to more nuanced scrutiny. There is more to the commons than the "tug of war between use value and exchange value," prompting the need to explore the evolving scenario of "practice and process"—a perspective that has even sought to replace the expression of "commons" with "commoning."[44] This is echoed in the suggestions that to understand commons, more so in transboundary contexts, it is crucial to go beyond merely tagging it with the question of resource management and to focus more on power relations, networks, and actors that facilitate the commoning process.[45] In other words, besides purely economic lenses to understand commons there are embodied and mundane entanglements between nature and human actors that deserve consideration in exploring this question.

The reasons for the quest for commons at the sites of the BISMRY rivers are not far to seek. These rivers originate in the Tibetan glacier region, preserving the largest concentration of fresh water outside the polar regions, earning itself the accolade of the "Third Pole." They collectively support nearly a third of the world's population with a rich range of options for livelihood, biodiversity, and transport networks spanning the sparsely populated high mountain ranges and the heaving plainlands around the Asian rims of the Indian and Pacific Oceans, presenting themselves as a life-nourishing "terrestrial ocean," as observed by Dan Smyer Yü.[46] As explored in the concluding chapter, the postcolonial era has seen these BISMRY commons face mounting existential threats, underscoring the urgent need to reassess their formative pasts.

V
Locating the BISMRY Rivers in Imperial and Environmental Histories

The centrality of rivers in contested imperial settings positions the book within both the ecological and imperial histories, without fully committing to either. Ecology in the sense of the fluvial architecture and geomorphological features of the BISMRY rivers is a significant focus of the book, as reflected in the first section of chapters 1 to 6. While acknowledging the distinctive ecological features of each river, the book takes exception to mainstream environmental history that privileges the declensionist narrative. The current planetary crisis that defines the Anthropocene is typically seen as the result of human interventions leading to progressive ecological deterioration. Was there no exception where nature was more intuitive than or at least at par with its human interlocutors? Responding to this query, *The Range of the River* tells the story of how Tibetan rivers themselves acted as formative agents, questioning the narrative of progressive decline. The rivers—with their physical flow, uncertain and seasonally variable velocity, rapids, rolling stones, silt, mud, and floods—were forces themselves that did not easily succumb to human advances. This perspective shows that global environmental decline did not occur simultaneously everywhere, and certain ecological zones exhibited resistance at least until the mid-twentieth century when the more aggressive acceleration of anthropogenic actions began to take their toll.

The rivers' power to counter the narrative of environmental decline has implications for imperial studies as well. Many imperial practices bore the mark of rivers' agency, not by design, but by compulsion. The empire sought to impose dominance around these rivers, yet rivers remained a force that shaped imperial practices along their own fluvial footprints through defiance or unnerving uncertainty, and the communication pathways they created. Despite the substantial ecological decline in the imperial era, ranging from deforestation to mining-related deterioration, navigable water spaces had to be preserved as arteries of trade and transport. In this context, rather than viewing the empire as an omnipotent force of destruction or a benign preserver of nature, the book suggests that the modern empires, like their premodern predecessors, harbored a symbiotic relationship between political power, production relations, commercial networks, and local riparian inhabitants and nonhuman forces like pack animals. These processes and trends, *The Range of the River* argues, contributed to the emergence of the river as a collective commons, defying political, economic, and geomorphological divides across regions and territorial sovereignties. The river was a process, a mobilizer, and a symbol of boundless life and longing—a fluid phenomenon that the empire was compelled to walk with rather than disrupt.

Thus the book explores the dynamics of relationship between the power of the river and the curtailed operability of the empire, giving rise to a new kind of political participation that defied geographical divisions of mountains, deltaic plains, and coasts. The empire met its more formidable successor in the form of the postcolonial state, which appeared far better equipped to curtail the range of the river. This book, therefore, suggests that—in the BISMRY region—ecology and empire informed each other, only to sustain a malleable third space.

Transregional mobility, flows, and assemblages in the BISMRY zones were not new, and Western empires sustained the premodern mobility practices. Although the British takeover of the North China Sea and the Bay of Bengal introduced an unprecedented shift, the British and other imperial forces became participants in historical dynamism rather than its determinants.[47] The entanglements between imperial agencies, ecological challenges and opportunities, and human resistance and collaboration at the site of the river led to common stakes in the rivers for all participants. In this context, the book argues for using multiple perspectives to explore the overlaps between the vibrant side of nature and a relatively subdued undercurrent of the empire, as far as the overwhelming presence of the river was concerned.

VI
Sources and Structure of the Book

The book primarily draws from the English-language archives. Each region, subregion, and locality within the vast area covered in the book encompasses too many languages to be grasped in a single monograph. Nevertheless, the diversity within the English-language archives and published materials has enabled me to go against the grain to understand both the imperial ambitions and frustrations concerning various river spaces, as well as the dynamics of local and regional politics, economy, and culture. A study of this scale would not have been possible without the imperial sources, as imperialists were precisely the ones searching for shorter and convenient trade routes through China, India, and mainland Southeast Asia. Since the British Empire had the widest presence in these regions, the British imperial archives and published primary materials provide the broadest window into interactions across this vast transregion. In addition, in some instances I have used French and German sources and English translations of local sources such as Vietnamese, Burmese, and Chinese, as well as sources in the original Bengali and Assamese. Conversations with local scholars during my visit to Bangladesh, China, India, and Vietnam also helped ensure that the predominant use of English-language sources did not detract from the local perspective.

In a view to explore the idiosyncrasies and outreaches of the BISMRY rivers, each chapter is devoted to an individual river, around which four key themes revolve: river network and nodal points of human and commodity flows, imperial encounters and their limits in the face of geomorphological challenges, autochthonous responses to imperial advances, and the affective nature of lived mobility and entanglements that constituted what one might term a "fluvial cosmos," which embodied forms of cohabitation with nature—a process that Heidegger called "dwelling." Covering six rivers across vast territories through four broad themes was an ambition I embraced with trepidation. But as the research progressed, I held on to the project as a pursuit of possibility rather than perfection. Such possibility of deriving meaning from studying multiple Tibetan-Himalayan rivers—from a methodological perspective—is eloquently expressed by Ruth Gamble and her colleagues in a recent pioneering work: "By following the multiple channels and temporal pulses of our research, we are using the river as method to find meaning in our multiple disciplinary meanderings."[48]

Following the exploration of the six rivers, the last chapter focuses on the mule in the physical settings of Yunnan and neighboring uplands, highlighting the aspects of intervalley connectivity with the caravans through the rugged mountains punctuating the BISMRY river networks. Individual treatment of the rivers captures local, regional, and transregional stories, while explorations of Yunnan crossroads from the vantage point of the mules allow a wider view of interlocking spaces of water and land that sustain the circularity of flows. There was something almost magnetic about the mutual attraction between the BISMRY rivers and the mules that played a central role in sustaining this connection. The concluding chapter briefly recaps how the BISMRY fluvial commoning was achieved through the fourfold historical processes outlined in each chapter. Finally, it discusses the postcolonial predicaments facing this commons.

Although the concepts of mobility, ethnicity, and commoning recur throughout the book, "empire" is retained in the title to emphasize the temporal specificity of the era under examination. This choice acts as a timestamp, capturing a historical moment that shaped the BISMRY rivers' trajectories between their precolonial and postcolonial career, marked by both continuity and disruption. At the same time, the river's sweeping range of entanglements with biodiversity and hydrosocial dynamics—both fluid and foundational—justify its claim to be an "empire" both as a metaphor and as a living embodiment. In this sense, "riverine history of empire" in the title holds a dual significance.

ONE

BRAHMAPUTRA
A Journey with the Son of God of the Universe

JAMES RENNELL, A PIONEER of British imperial cartography, was tasked in 1765 with surveying the Ganga River to facilitate commodity flows from eastern Bengal to Kolkata Port. During the survey, he discovered that to the east of the Ganga, a larger river called Brahmaputra was flowing into the Bay of Bengal and that its northernmost part was within two hundred miles of the Chinese province of Yunnan.[1] The Brahmaputra's physical proximity to China was surprising to Rennell as he was the first European of any authority to discover this river and its transregional outreach. However, it was already a major trade and communication route between Tibet, China, and northern Burma.[2] In the nearly two centuries that followed, the British strove to establish imperial authority on the navigable course of the river and also used it as a terminal to reach Tibet and China, triggering a vivid transregional itinerary.

Recent historiography of the Brahmaputra has privileged its more cartographically visible spaces than its actual transregional propensities resulting in three spatially delimited historical narratives. Since the late colonial period, a territorially sensitive nationalism on the Ganga-Brahmaputra delta paved the way for the emergence of East Pakistan and later Bangladesh. The Bengali national motto became "Tomar amar thikana, Padda-Meghna-Jamuna" (Our collective destination: Padma-Meghna-Brahmaputra), which had little reference to the rest of the Brahmaputra upstream.[3] The delta was a national space first, and an ecological space second. In the middle to upper Brahmaputra Valley, the river acted as the fluvial nerve of Assam's regional imagination. Assamese self-assertion, through a dual critique of imperial exploitation around the Brahmaputra Valley and the postcolonial central government's troubled encounters, has been at the core of historical literature on Assam.[4] On the northern edges of the Brahmaputra, a rich range of literature has emerged that may be clustered as borderland studies. This literature has put forward an impressive critique of territorial nationalism and postcolonial state maneuvers at its margin, and fostered a sense of cultural and political affinity that synced more with the highland ethos of the Southeast Asian Massif than with the lower Brahmaputra basin.[5]

All three strands of studies of the territories along the Brahmaputra basin are tempting invitations to explore the extent of connectivity across various geomorphological spots of the river system, from Tibet and the Southeast Asian Massif down to the Bay of Bengal. The cartographic coup that marked the partition of India in 1947 largely diminished the flows of transregional mobility and connectivity that the British Empire had sought to facilitate. Imperial schemes had aimed to exploit the Brahmaputra's ecological resources with mixed success, but in the process the river sustained the precolonial connections that bounded the Bengal Delta, much of northeastern India, southeastern Tibet, northwestern Burma, and southwestern China. As a small number of emerging literature shows, transregional and transborder flows become more intelligible with the rereading of the history of mobility and connectivity via the Brahmaputra and its neighboring river network (Figure 1.1).[6]

I

James Rennell had described the Brahmaputra as an "arm of the sea." From the vantage point of the Bay of Bengal, Rennell's description would seem quite fitting, as the river emerged from the watery labyrinth of the world's largest delta and majestically revealed itself above the Ganga and the Meghna. But if gazed at from a point in the eastern Himalayas, the Brahmaputra would look like a child of the mountains. It was more a collection of the flows that trickled down the numerous tributaries and subtributaries from the eastern Himalayan ranges. In 1838, Montgomery Martin, a British official and author, found fifty-eight rivers forming the fluvial network of the Brahmaputra, and he thought that China could be reached by Assam by water.[7]

On the way to the Bay of Bengal on a course of about 1,100 kilometers, tributaries from the north above Sadiya and on both its banks emerge as spokes of an umbrella. On its right bank, it is joined at regular intervals by the Tsangpo (Siang/Dihong), Subansiri, Kameng, Beki, Manas, Jaldhaka, and Teesta, and

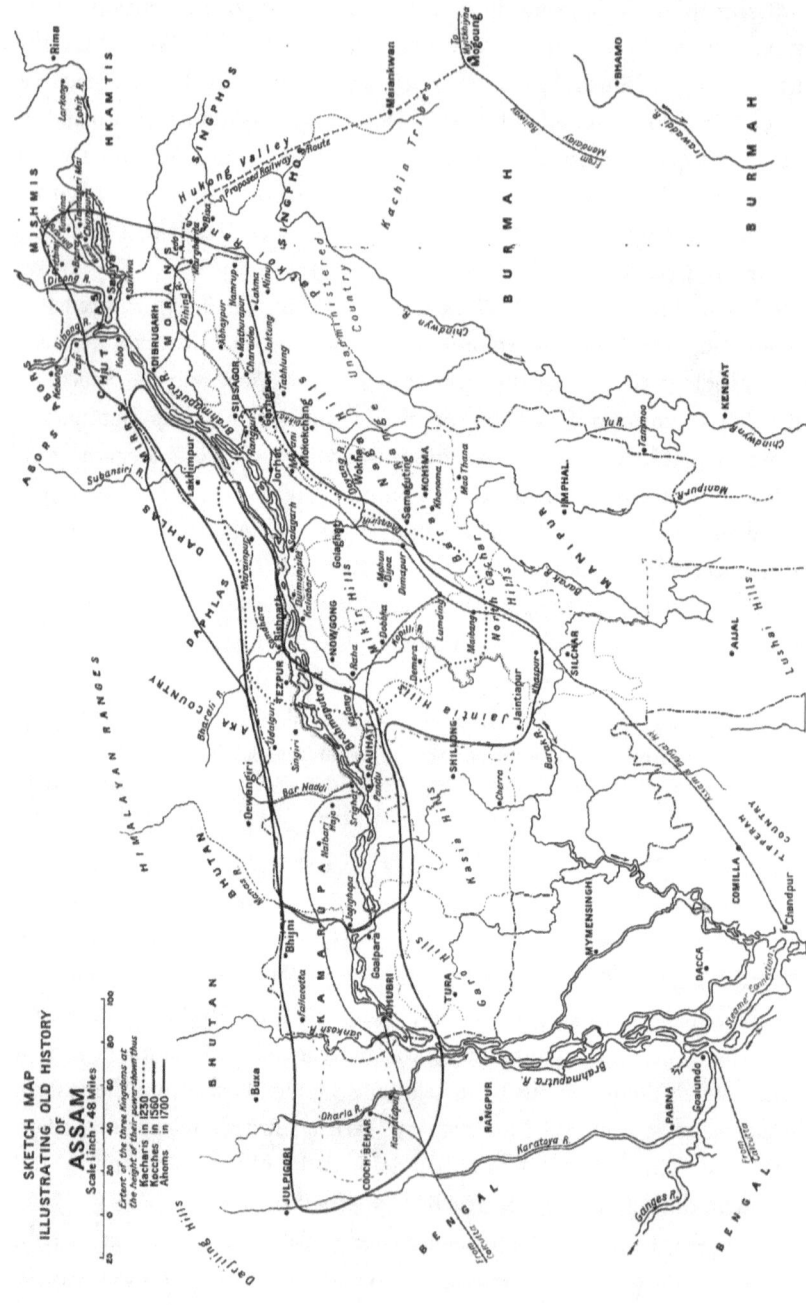

FIGURE 1.1: The Brahmaputra's outreaches to eastern Tibet, Burma, and Yunnan via Lohit River. Source: L.W. Shakespear, *History of Upper Assam, Upper Burmah and North-Eastern Frontier* (Macmillan and Co., Limited, 1914).

further down by the gorgeous Ganga. On its left bank, it is joined by the Lohit, Noadihang (joining from the northern borders of Burma), Bodhi Dehing, Disang, Kulsi, Mornoi, and the mighty Meghna further down the delta.[8] Each of these tributaries in turn is enriched by numerous smaller snow-fed streams from upper valleys. Within its length of about 120 miles, for example, the Lohit drew more than a hundred small rivers, rivulets, gullies (nullahs), and other forms of water channels that eventually joined the main body of the Brahmaputra.[9] Further east beyond its network above Sadiya, the Lohit stood out as a fluid outreach of the Brahmaputra further east. To Kingdon Ward, the Lohit was "particularly interesting" because its upper reaches flowed close to the Salween and, in its middle course, it flowed equally close to the Irrawaddy, although divided by a narrow range of mountains.[10]

Within this dense fluvial system, the flow of commodities and trade was shaped by several nodal points. From the deltaic side, one major communicational hub was Goalundo, where the Ganga met the Brahmaputra (Figure 1.2).

FIGURE 1.2: Boats at Goalunda, the terminus of the Great Eastern Railway at the confluence of the Ganges and Brahmaputra Rivers, 1895. Source: William Henry Jackson. https://www.loc.gov/resource/wtc.4a02672/. Library of Congress.

The confluence of the two Asian titans here meant that commodity flows from the entire Ganga Valley found a way to both the upper Brahmaputra and down to the Bay of Bengal. From the mid-nineteenth century, the development of tea plantations in Assam, the discovery of petroleum, and strategic considerations in the Tibetan and northern Burmese frontiers prompted the British to find ways to connect to the upper Brahmaputra via Goalundo and Sirajganj, located a few kilometers upstream. By the turn of the nineteenth century, between Sirajganj and the Bay of Bengal ports, thirty thousand boatmen facilitated trades worth £2 million sterling. By 1870, the annual value of commercial transactions in Sirajganj was at least £3 million sterling, more so unofficially—a commercial bonanza, that was reflected in about fifty thousand boats registered in Sirajganj in 1877.[11] Unofficial data would have it at a larger figure. Bengal Viceroy Richard Temple on his visit there described the bursting of a "floating city" where "tens of thousands of boatmen, workmen, and traders are congregated," and where "boats of all sizes in thousands are moored and lashed together, thus constituting stages, almost roadways along which people can move to and fro."[12] Branches of six European firms and banks including the Bank of Bengal had an active presence in the town.

If Sirajganj, along with its own regional trade in jute and rice, was a vibrant transit point in the middle of the Brahmaputra connecting the ports on the Bay of Bengal, a significant part of that vivacity originated in three towns further upstream: Guwahati, Dibrugarh, and Sadiya. While Dibrugarh was the source of the regional trade of tea, Sadiya catered to the Tibetan, Chinese, and Burmese trade flows. It was located just above the confluence of numerous rivers that formed the Brahmaputra, many of which acted as trade routes from southern Tibet to the Brahmaputra Valley. These tributaries included the Kuru (Manas in Assam), Subansiri, and Nyamjang in Tawang, which was connected to a trade route with Lhasa via the valley of the Manas.[13] Another important route to connect Sadiya to Tibet was via Tezu. The Bhutanese often joined with the Tibetans to come here to procure tea and package it in Batang from where "thousands of yaks and mules annually crossed the Himalayas, on their way to Lassa."[14] While the Brahmaputra was connected with Lhasa via Tawang on the Manas River in the west, in the north it was connected to the commercial site of Miri Padam in the Abor district around the Dihang River.

A prominent trading post that connected India and China via Tibet was Rima, about 130 miles northeast of Sadiya, and located on the upper end of the Lohit.[15] Rima connected Sadiya with Walong, Batang, and finally Lijiang in Yunnan on the Yangzi.[16] These intervalley connections were possible due to well-trodden passes through the Brahmaputra-Salween divide, one being spotted by Ronald Kaulback, a British explorer and botanist, during his trip along the upper Salween.[17] If the Lohit paved traffic between the Brahmaputra and Yangzi across the Mekong and Salween, it also clearly offered a line of communication with the Irrawaddy across Khampti, which in turn retained communication with both Kolkata and Mandalay. At the Khampti capital, Prince d'Orléans, a young and enthusiastic French explorer, met a group of Gurkhas who had arrived there after traveling from Darjeeling to Kolkata, and then to Rangoon, Moulmein, Mandalay, Bhamo, and Mogaung, showing a circular mobility of people and commodities along the Brahmaputra–Bay of Bengal–Irrawaddy–Salween route.[18] It seems if the Brahmaputra's one arm went to central Tibet to the west, another went to Yunnan via Lohit to the northeast, and the other toward the tributaries of the Irrawaddy via Khampti in the east.

Sadiya continued as a commercial bridgehead between the Brahmaputra Valley and Tibet, and there were no major disruptions in the existing trade patterns despite seasonal natural hardship and occasional political upheavals. Tibet mainly exported wool, yak tail, and live animals and imported cotton piece goods, metal and metalware, woolen piece goods, and silk products.[19] This part of the Tibetan plateau was compared by Markham to the Collao of Peru, lying between the maritime zone and eastern cordilleras of the Andes, which was a hub of water communication at twelve thousand feet above the sea.[20] Markham's comparison followed his observation of exchanges in Assam fairs dealing with Tibetan gold dust, salt, musk, cow tails, woolen products, and horses, in exchange for lac, madder silk, clothes, and dried fish. From these river points, Gosains or trading pilgrims from India departed for Tibet in "large numbers," and their trade dealt with "articles of great value and small bulk," and they traveled "without ostension and often by paths unfrequented by other merchants." Using a combination of river and land routes, the Bhutias also brought the products of Bengal and Assam to Lhasa.[21]

Most of the trade that flowed from Tibet further down the Brahmaputra went to the Kolkata port via Sirajganj and Goalundo, constituting a total annual trade of about 3.2 million rupees in the early twentieth century. These statistics do not include Assam's trade with Bhutan and several ethnic groups, especially Abors and Mishmis, with whom trade was considered "even larger."[22] Of the total exports from Kolkata, 56 percent were made up of raw jute and jute products, 11 percent constituted tea from Assam, and the rest were a combination of trade flows along the Ganga and Brahmaputra Valley, in which the amount of Tibetan and Chinese products must have been considerable.

II

It did not take long for the British to appreciate the commercial merit of the Brahmaputra transregional network. These interests were reflected in the explorations and expectations of various routes. The Brahmaputra-Yangzi route via the Lohit spawned an imperial desire for connections between India, Tibet, and China. Arthur Cotton, an influential irrigation engineer, thought that although it was not possible to establish a thorough water transit between India and China, a combination of water and land routes of less than 250 miles was worth the expenditure. For Cotton, the ideal route was between Sadiya and Lijiang, on the westernmost navigable point of the Yangzi (locally known as Jinsha). Cotton recognized it not only as the shortest route between the "hearts" of India and China but also as a vital link between the two major water transit systems.[23] He considered the Brahmaputra the most suitable route to China, especially when compared to other transregional connections, such as the 5,000-mile link between St. Petersburg and Nanking, or the 1,200-mile route between the Indus and the Caspian Sea. In the 1860s, this route was also favored because northern Burma was still under the Burmese king, complicating a direct route from British Burma to China. In addition, Cotton noted that after more than 1,000 miles of a river trip on the Irrawaddy, the connection was only made with southern Yunnan, not the Yangzi Valley. The idea of the Brahmaputra-Yangzi's connection in the imperial circle was also based on the possibility of commercial flows between Shanghai and Kolkata. There were even suggestions to construct a canal to join these two river valleys.[24] The argument in favor of Brahmaputra-Yangzi's direct connections was further strengthened by the pressure that the British felt from other competing imperial powers in Shanghai, necessitating that they found a westbound outlet for Anglo-Chinese trade.[25]

Thomas Cooper, an avid British explorer and the merchant tasked with finding convenient trade routes across India and China via Assam and Tibet, agreed with Cotton in favoring a route between Brahmaputra and Yangzi Valleys. Cooper eyed the route that started from Sadiya and crossed over to Zayu, eighty miles from Sadiya on the Assam-Tibet border and then to Batang, in a view to connect the "grand artery of Central Asia." Cooper observed that

Tibet imported between four million and six million pounds of brick tea from China annually, and that this route could facilitate the replacement of part of that Chinese trade with Assam tea, generating gold and silver in return. He also hoped that central Asia and Russia would find ways to conduct trade on this route via Tibet to gain access to the Assam market, while Manchester goods would find a way to central Asia. Cooper discounted the Rangoon-Yunnan route due to the improbability of Sichuan trade making its way to Burma via Yunnan, since its natural route was to Shanghai via Chongqing and Hankow at a much lesser cost than the Chongqing-Bhamo route via Dali.[26]

The railways initially appeared as a key imperial tool for transregional connections. The first railway line in the 1850s from Kolkata to east Bengal in Kushtia on the Ganga was conceived as the starter for a line between Kolkata to Canton. The railways initially appeared as a formidable competitor to the Brahmaputra water transport network, but made little impact in the long run, compared with the persistence of the waterways. The railways reached Assam as late as 1883 and were not extended to upper Assam until the 1890s, with no railways at all above Dibrugarh. People continued to be dependent on water transport, while the data on railway commuters showed a decrease of 55 percent as late as 1911.[27] Steamers brought a remarkable change in the Brahmaputra transport network, stretching the navigability of the Brahmaputra considerably, yet they failed to break the rhythm of the conventional transport made up of indigenous boats. Saikia notes that nearly six hundred boats annually traveled upstream from Sirajganj and Dhaka for transporting timber from Assam "even during the heyday of the steamer."[28] With the freight charge set higher for the Kolkata-Assam route than for the Kolkata-London route, the steamers failed to displace the boats—a scenario that lasted well after decolonization. While the Brahmaputra couldn't be conquered by the modern technology of railway and steamers, the indigenous boats together pushed the British Empire's transregional quest a little further and continued into the twentieth century.

A remarkable example of high imperial efforts in spanning China, Assam, Burma, Bengal, and the Indian Ocean was Nathaniel Curzon's establishment of the province of Eastern Bengal and Assam, with the Brahmaputra serving as its vital artery.[29] It was within this larger spatial rearrangement that Williams, a locally posted official, declared in 1908 that the physical barrier between the Lohit Valley and Yangzi Valley was not unsurmountable. He suggested that after the actual rise to 5,000 feet, there was no higher altitude to cross and that Lohit's banks appeared "specially formed for a road" with a gradual ascent of 1,900 feet in 70 miles, being "a natural highway into Tibet." Williams's assessment of potential connectivity between the Brahmaputra and the Yangzi across the Mekong and the Salween was informed by his interest in the possible construction of a railway that would save time and distance between Kolkata and Shanghai. He agreed that such connections were visionary, but "not more visionary than twenty-five years ago was that of a modern hotel at the Victoria Falls of the Zambezi."[30]

Another British official contested the idea that difficulties in communication between Assam and China were due to the "convolution" of the Himalayas with the Patkoi Range. He didn't observe any such convolutions of mountains as he reached from Rima to the left bank of the Brahmaputra in four days, Rima being also four days away from Batang on the Yangzi. The Brahmaputra's proximity to Rima meant that it was also connected to Lhasa and Beijing, because from Rima couriers left daily for those places.[31] As another nodal point between the Brahmaputra and the Yangzi, Batang too offered considerable advantages. Cooper noted that traders from Mongolia and China visited Batang to purchase tea. Besides being a trading post itself, the fertile valley of Batang produced wheat, peas, turnip-flavored vegetables, cucumbers, Chinese cabbage, potatoes, leeks, pears, peaches, walnuts, and watermelons, while fish, mutton, and chickens were plentiful. There was enough fodder supply for the mules and other pack animals, which stood up "to their knees in fresh-cut green wheat."[32]

In the course of World War II, the Batang route attracted renewed imperial attention. The proposal to connect Sadiya and Lijiang was revived by the construction of a motor road across Si La on the Salween-Mekong Divide and the Mekong to Weisi. The hope was that this route would divert Tibet's wool trade to Sadiya, rather than the longer route through Lhasa or to the Yellow Sea via the Huang He. Following Cotton and Cooper, British botanist Kingdon Ward had already suggested in the 1930s that the Rima route was the "shortest and on the whole the best overland route between India and Western China." The Chinese Tai Kyi Company, for example, found several advantages along the route: reduced freight cost, less likelihood of banditry on the way, and "fewer hands to grease over a shorter distance."[33]

These imperial hopes of connecting India and China through the valleys of the Brahmaputra and Yangzi, however, met with little success on the ground. Despite a century of planning, surveys, and efforts at the height of British imperial supremacy, the various proposed routes between the Brahmaputra and the Yangzi via Sadiya and Lijiang never materialized.

As missionary E. H. Higgs suggested, the Chinese fear of losing the tea trade and the Lama's apprehension about the Tibetan conversion to Christianity made the imperial efforts of connectivity from Sadiya to Batang seem like chasing "a chimera of a distorted imagination."[34] Another factor that undermined this route was the British annexation of northern Burma in 1885, which turned attention to the possibility of using the Irrawaddy-Yunnan route ushering a more direct connection to China than via Tibet and the north of Yunnan. The interest that was raised in this route during World War II also soon disappeared. The projected Rima motor route from Sadiya to Lijiang, conceived as the "northern route," was discarded because of the difficult and rugged landscape between the river valleys it had to cross.[35] Two other projects were also mooted, but nothing worked practically and all hopes dimmed with the new realities facing postcolonial India and China.

Although attempts to secure routes for tapping and controlling maximum flows of trade between the Brahmaputra and Yangzi rivers failed despite imperial maneuvers, it is noteworthy that trade continued nonetheless, as previously mentioned. How can we understand the continuity of transregional trade mobilities across Yunnan, Tibet, Burma, Assam, and Bengal along the Brahmaputra basin when the imperial schemes of highways and railways failed? The answer lies in how the riparian inhabitants maintained and utilized a locally connected mobility network. This flow of goods along the Brahmaputra's nodal points and connections across Bengal, Assam, Tibet, Yunnan, and Burma occurred despite, rather than because of, British imperial efforts. The empire did facilitate the flow of goods to port cities from inland, but this was only feasible through engagement and compromise with several ethnic groups who effectively controlled the routes connecting the upper Brahmaputra networks. Examining these imperial-local encounters is crucial for understanding the continuing mobility of people and commodities on the Brahmaputra from areas beyond its navigable reach.

If the empire was a process rather than a structure, as David Ludden has suggested, then this process was contingent on other processes, including those sustained by local inhabitants around the river and river-inclined nodal points of the trade.[36] This explains why the commercial mobilities were neither significantly enhanced nor remarkably destabilized because of the British failures to connect the Brahmaputra and Yangzi corridors. Despite occasional natural disasters and political highs and lows, there was business as usual over a longer period. The following section discusses how the human and commodity flows beyond the imperial grids were negotiated by the local inhabitants at the sites of the Brahmaputra and beyond.

III

Powell Millington, a British military commander serving in Assam, once quipped: "When the argument is imperial the 'spot' that matters is not a ten-acre clearing in a valley of Dihong, but a secretariat in Simla, or an office in Downing Street."[37] Recent literature is replete with suggestions that it was neither the colonial field office nor an imperial metropolis where effective control emanated from. Political contingencies and multiple layers of temporality informed the colonial power to find stable footholds in the territories by engaging local ethnic communities living across Indo-China border areas, which Guyot-Réchard terms "Shadow States."[38] Despite the ethnographic other that the British administration constructed out of the various ethnic groups in the Assam-Tibet-Burma borderlands and despite occasional military expeditions launched against them, the administration's relationship with these groups was of uneasy engagements rather than ceaseless hostilities. This was because the more the British Empire became knowledgeable of these borderlands, the more convinced they became of the need for collaboration with them as intermediary transactional partners. The political relations around the headwaters of the Brahmaputra could be better understood in this context.

The ethnic groups who inhabited the headwaters of the Brahmaputra, from the west to the east, included the Akhas, Duphlas, Abors, and Mishmis, among many others. A colonial official once bragged that these "untutored savages must first be taught that for the unseen power which resides at Gowhatti [Guwahati], or Calcutta, or Dacca no river is too deep to be crossed, no jungle is impenetrable, no stockade is so beset with snares and spikes that it cannot be stormed" with mountain guns and Gurkhas. But he was also quick to suggest that the government of India "will have to lay down definite rules of some elasticity suited to the requirements of each tribe."[39] This duality of imperial approach to the borderland ethnic groups ran deep primarily because the colonial state realized that "nothing can be achieved in these directions so long as the native tribes maintain their present hostility to all commercial intercourse."[40] This imperial desperation and near resignation to the "native tribes" left a deep impression on the agency of these people who maintained the flow of commodities and people along the Brahmaputra and beyond its navigable points in Sadiya, while the high hopes to establish firm imperial networks were dashed. As Bodhisattva Kar has perceptively shown, the British assisted the "nomadism of capital" in a way that breaks the perceived dichotomies between the homogeneity of colonial capital and resistance of the indigenous groups.[41] This was because, for the capital to flow barrier-free, the mobilizing force of these ethnic groups in a fluid and rugged landscape was essential to the empire.

The Abors, a representative example of the trend, inhabited the basins of both the Dihong (Siang) and Dibong Rivers, which joined the Brahmaputra from the west and north and connected China-Tibet routes to and from Lhasa. They practically controlled the river network that lay between the loops of the great bend of the Tsangpo to secure their habitat, livelihood, and trade (Figure 1.3).[42] The Abors grew cotton, tobacco, opium, bamboo, plantains, and lemons, and the Abor Hills were home to many herbal medicines, such as aconite, that they exchanged with the Tibetans for brass pipes, beads, copper cooking pans, silver ornaments, salt, and yaks. Frederick Bailey, a British official, crossed one village in the Tibetan-Abor border where he met a man who told him that his grandfather had migrated there from Bhutan and they traded salt and rice with the Abors for cotton.[43] The Abors' exposure to the crossroads of the flows of Tibetan, Chinese, Indian, and British commodities was embodied in what they dressed up with: hats ornamented with yak tail, woolen cloaks, steel pipes and tobacco, long Tibetan knives, and arrows with aconite.[44]

E. H. Higgs, the missionary who lived among the Abors between 1854 and 1856, recalls that he found them "commonly" talking about their communication with China. Although he was not certain if the Abors themselves traveled to China, he found almost all the ornaments worn by them were of Chinese origin, especially a peculiar blue bead, a bell-metal pipe stamped with Chinese characters. He was able to verify the Chinese origin of items there with the help of a Chinese tea maker, who had been living there for many years. Higgs was convinced that these articles had come into the Abor region direct from the Chinese, who often exchanged salt with the Abors.[45] But the Abors' main commercial connections were with the Tibetans. The Tibetan resistance to both Indian tea and European merchants across the Assam border reflected this strategic bonding, as the Abors, who had established trade routes along the Brahmaputra tributaries, primarily served Tibetan needs.

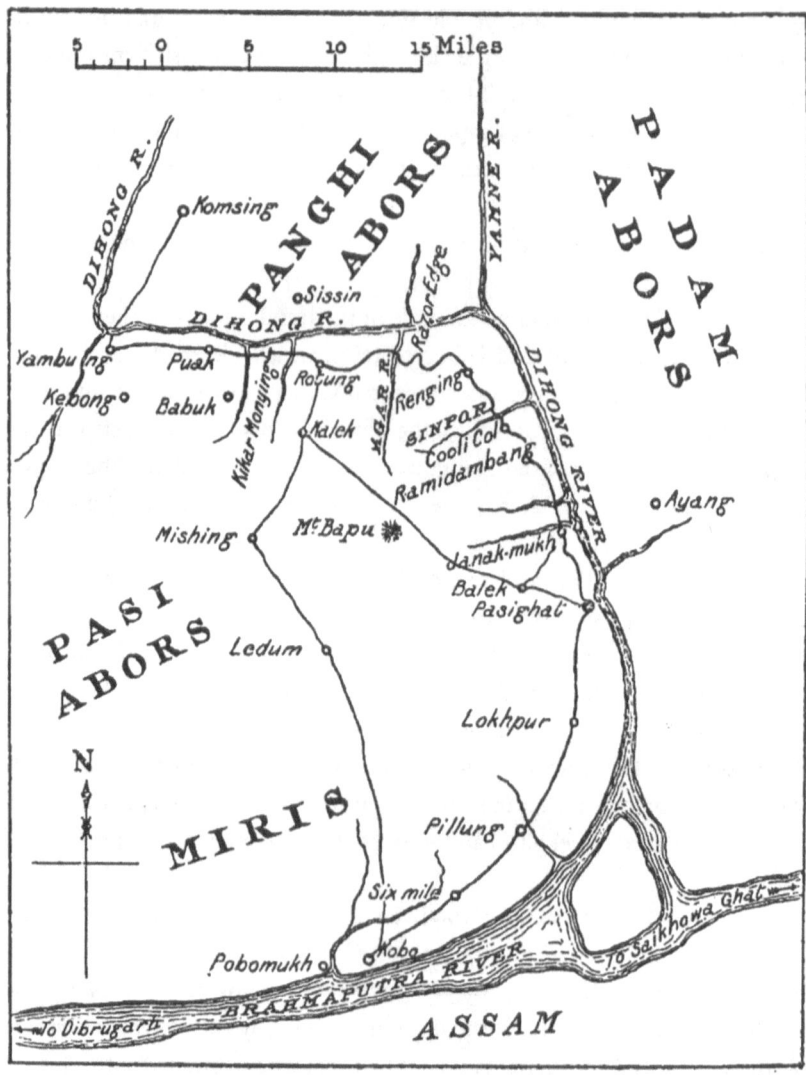

FIGURE 1.3: Abor settlements on the upper Brahmaputra basin. Source: Powell Millington, *On the Track of the Abor* (Smith, Elder & Co., 1912).

Tibet's refusal to allow any foreigners into their territories was partly due to perceived ecological threats. They argued that tutelary deities were offended following the arrival of two Englishmen in 1846. Since then, "year after year the people were afflicted with various sicknesses, the horses and cattle were struck with epidemics, the land was ravaged by locusts, the crops were deficient and the country in many ways suffered injury."⁴⁶ The aversion to colonial travelers is also reflected in the narrative of an English official, whose Tibetan assistant Wang Shu told him not to shoot or even look at a particular hill with field glasses, with a warning that a European who did so earlier died and that his grave nearby was a proof of his death.⁴⁷ These conversations well might be an example of passive resistance to protect nature, but it was also possibly due to the fear of conversion to Christianity and the fear of loss of profit from the tea trade with China, worth about £4 million to £6 million, that made the Abors a buffer and intermediary ally.⁴⁸ The Tibetans, who "reck little of what we call 'natural' frontier," as Kingdon Ward noted, also came down to Sadiya with their trade items.⁴⁹ From the Abor side, their trade with the Tibetans was "too important to allow them to go against their neighbors."⁵⁰ The Abor visits to the Tibetan villages on the Tsangpo above the great gorge generated a "backward and forward surge of the tribes trapped in the hills between the plateau and the plains; those living just under the plateau ascending to the plateau, and those on the fringe of the plains descending to the plains, the two parties mingling at other seasons."⁵¹ This India-Tibet-China trade route exploited by the Abor-Tibetan alliance inevitably attracted the covetous attention of British imperial forces.

The first British-Abor hostility began in 1848, resulting in at least six subsequent expeditions until 1912. The first one was caused by the colonial infringement of Abor rights on fish and gold dust available in their hill streams.⁵² Soon the conflicts flared around trade routes lying along both sides of the Dihang and other rivers and connecting the two prominent villages of Shimong and Karo. This "broadish, well-defined trade path" was trodden by "hundreds of laden yak bringing commodities to Shimong, whose inhabitants distributed the same throughout the northern Abor clans." This and other subsequent conflicts generally ended with the Abors being left alone to deal with their own trade network. The expedition of 1912 resulted in the British assurance of allowing Abor access to trade facilities at Rottung, especially along the riverbank paths that connected the trade routes or salt wells. The British administration also assured that they would not take away any of the Abors' land, villages, or fields and that they would not interfere in matters which the village council had settled.⁵³

During a theodolite survey in 1913, officials discussed with the headmen of different Abor groups including Anaia, Kano, Amili, and Mrambon the need to keep open the road to Sadiya for all without any interference, for which he received their commitment.⁵⁴ Since the British were unable to establish a secure commercial relationship with the Tibetans, the best policy left for them was to remain engaged with the Abors for an indirect connection to Tibet. The policy of noninterference with the Abors was shaped by many factors, but the central issue of keeping the water-land trade network between the Tibetan frontiers and the Brahmaputra remained constant. In the run-up to an expedition against the Abors in 1930, British officials decided to maintain the status quo, more so owing to the financial constraints due to the global economic recession. The administration in Delhi reaffirmed that "no general scheme of penetration nor any definite forward policy in the Abor Tracts has ever been sanctioned" and it advised keeping things within the "compass of that policy."⁵⁵ This policy of noninterference was complemented by an appeasement policy that had included, since 1862, the continuing of the Chinese tradition of offering *posa* (gifts) to the inhabitants of border areas. Initially, they were given the *posa* of iron hoes, salt, rum, opium, and tobacco, and later this was turned into an annual stipend of 3,400 rupees. L. W. Shakespear, an administrator and author, sighed that the "Abors after all these futile efforts at punishment on our part and their recent substantial gain should have had an exaggerated notion of their own powers. Their outrages in various petty ways still continued, and still they received their 'Posa'!"⁵⁶

If the Abors facilitated the trade flows between eastern Tibet and India, the Mishmis had nurtured connections with eastern Tibet as well as Yunnan with the Rima, Batang, and upper Irrawaddy regions as nodal points.⁵⁷ The Mishmis, including the Idu, Miju, Digaru, and Deng clans, predominantly inhabited the riverine network east of the Abor territories, extending from Dibang to the Lohit-Zayu basins. Sourced in the Zayul Province in southeastern Tibet, the Lohit cut through the eastern leg of the Himalayan range. Before reaching out to the Brahmaputra at Sadiya, it ran for about 120 miles at 17,000 feet above sea level, while receiving the Ghalum River in its southernmost bend in Mizong near the Burma frontier.⁵⁸ The Mishmis living along the valley of Lohit-Zayu thus found themselves at the crossroads of flows from northeastern Assam, extreme southeastern Tibet, and northwestern Burma. They stood their ground and used those watery and hilly territories to secure their bases on the trade routes between India and China. Often envisioning China as their ancestral homeland while residing closer to Indian territories, they straddled both worlds, drawing from the influences of each.⁵⁹

Descending at a rate of 40 feet a mile, the Lohit River offered limited navigation, although most parts upward from Sadiya were "comparatively placid" with a width of 150 feet.[60] But pathways along the river valley and the Mishmi's fine bridge-making skills meant that communication with Sadiya at the Brahmaputra remained intact. Trade between the Mishmis and the Tibetans was part of everyday life, and a passing tourist in the region would not miss small groups of merchants scaling the trade routes. Frederick Bailey, on the last leg of his trip from Beijing to Sadiya, met four Mishmis trading with the Tibetans and carrying 436 bags of grain, four rolls of Assam silk, four musk pods, and small swords, among other items.[61] The Mishmis themselves were consumers of Chinese products, as reflected in their wearing of Chinese beads and bell rings, bracelets of silver and brass, and Chinese-manufactured smoking pipes—most of these commodities being procured from the Tibetans.[62] The Mishmis exchanged *teta*, a kind of febrifuge, and musk for yaks, knives, spearheads, iron, iron cooking pots, beads, and brass pipes of Chinese origin. The items they exchanged with the Tibetans were brought to Sadiya and exchanged for cotton items, blankets, *burra cupras* (large pieces of clothes 12 by 2 yards made in Lower Assam), and rhea fiber.[63]

Sadiya was thus important to Abors, the Mishmis, and the British administration as a vital node for transregional trade and communication. Since the Abors and Mishmis forged an alliance, if loose, with the Tibetans, there was little chance for the British to take hold of the trade routes connecting Tibet and China. Thomas Cooper's experience is representative of the situation. As he reached the Tibetan border from Assam to cross over to Tibet, the local chief of the Lama clan, Samsung, requested him to turn back. He handed over to Cooper a letter from the "Chinese Rajah" of Rima along with a gift of a Tibetan dog and a knife, with the advice that if he didn't go back with these gifts and moved ahead across the border, he would do so at the risk of "encountering many such knives and watchdogs."[64] The Meju Mishmi chiefs accompanying him also agreed that he shouldn't proceed further. This was a double blow to Cooper, because a year before he had attempted to cross from the Tibetan side over to Assam and was stopped at the border. His mission to launch an imperial trade route between Assam and China through Batang thus met with a failure, and the only option for Cooper was to make sure that Sadiya remained open and accessible to all trading parties. Cooper's failure was translated into the consolation that he was at least able to bring to Sadiya the Mishmis who guarded transborder trade routes and thus could be "led to trade with Assam in the plains, and become medium of introducing tea into Thibet." He therefore "begged" them not to stay away from Sadiya.[65] This practice of appeasement continued until the end of the British Empire in India, as seen in the confession of an official during World War II that part of their duties was to make friendly contacts with the Mishmis.[66]

Translocal mobility was established in topographical setting that the imperial forces were unable to grapple with, which explains the frequent failure of colonial expeditions against them. Little wonder that after the last military post at the forty-eighth mile north of Sadiya, all territories were left to the locals. Even within the proximity of military outposts, it was difficult for the soldiers to find an easy way around. Captain Cross, who was in charge of a survey, noted that for army personnel a notebook and a pencil often became a burden, and more so when equipped with a prismatic compass, clinometer, a .45 pistol, thirty-six rounds of ammunition, a camera, and haversack.[67] On top of this, the road network that the government officials and the army wanted to build was different from those of the Mishmis, being liable to be washed away by floods and landslides, and requiring government forces to use the Mishmi tracts, which were "almost vertical," but not susceptible to landslides.[68] While imperial reconnaissance groups struggled in these routes, Mishmi and Tibetan traders comfortably used them. Referring back to Cooper's expedition, a Khampti village chief of Shang Kam village, who was a visitor to both Bengal and China, quipped that Cooper "took thirty days accomplishing a distance [across the Assam and Tibet border] which our people can walk in four days."[69]

The commercial and political space that the Mishmis, among others, nourished to facilitate commodity flows between the river basins was maintained by intelligent utilization of social relations, the material world, and the technological skills available to them. Contrary to some anthropologists' suggestion that the ethnic groups living in this region lived by revenues exacting from the trade that crossed their territories, trade was embedded in the daily life of the Mishmis, so much so that it took center stage in the marriage and family life. As Cooper noted, young Mishmi men were supposed to save enough to make a house of their own to bring their wives after marriage. Such custom was a "great stimulus to the young men in their musk-hunting and trading excursions, for until they pay for their wives they hold no position, and their wives and children have to work for the benefit of the wife's family."[70]

Keeping communicational routes alive was also necessary for the Mishmis to facilitate the mobility that took place in their territories at the crossroads of transregional trade. During the survey of the Dibong River areas, a colonial official at Sadiya met with a fine bridge built by the Mishmis in a few days, which was ninety yards long and above the highest flood level, and strong enough to last about five years.[71] In some places, there were biodegradable rope bridges with wooden sliders, while in others temporary bamboo bridges were used, which were good through the winter, and weak or washed away by summer flooding and repaired or replaced after the rainy season.[72] The cane bridge on the forty-yard-wide Irbung River was the best that a colonial official had ever seen, on which sepoys carrying long rifles and dogs all were effortlessly mobile.[73] All this meant that, despite a general lack of mobility during the winter season due to heavy snow, the Mishmis were able to keep mobile if needed. For example, along the Delei River Valley, a northern tributary of Lohit was crossed by the Mishmis right through snowy December.[74] In the Abor-Tibet border region, below the Tsangpo bend, Bailey reported many trade routes running along Brahmaputra's tributaries where he encountered bridges of "huge wooden piles" and boats "rather better made than most I have seen, of leather stretched on bent willow about 7 feet by 4."[75]

The trade flows and the stronghold that the Abors, Mishmis, and Tibetans developed around the headwaters of the Brahmaputra were sustained by

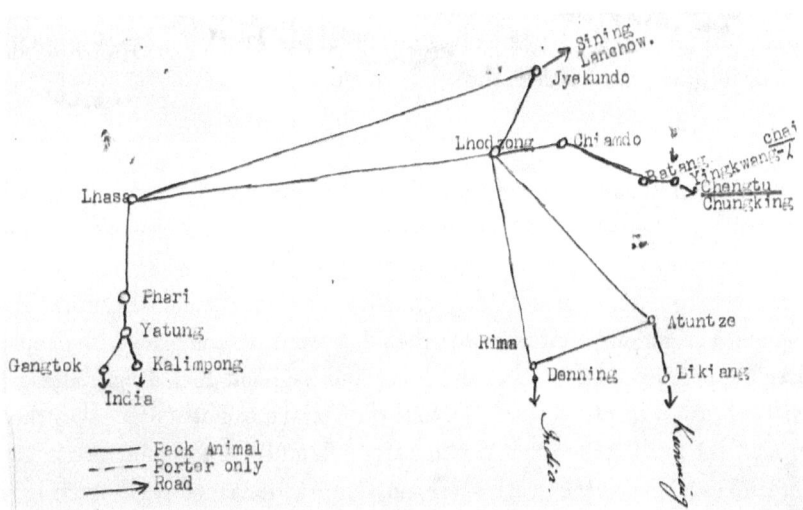

FIGURE 1.4: A hand-drawn diagram of Brahmaputra-Yangzi trade routes connecting India, China, and Tibet. Source: British Library, IOR/M/4/2404: Proposed Survey of upper Salween and Tibet, 1944.

a multiethnic network that stretched as far as Lijiang on the Yangzi, north of Dali. Lijiang was an important nodal point for connections across Yunnan, Sichuan, and the longer Yangzi trail all the way to Shanghai (Figure 1.4). These connections were more often mediated by state mechanisms that used the tributary system holding intervalley mobility together. As Prince Henri d'Orléans noted about a kingdom around Atuntze, located south of Batang and north of Dali and about the midpoint between Sadiya and Lijiang, the chief was of "noble blood" belonging to the ancient royal family of Lijiang. The kingdom technically stretched to the Yangzi to the east and the Irrawaddy across the Mekong and Salween in the west, but the Chinese court in Yunnan did not allow a free hand for their chief in all areas, to counter possible political threats. So, within this region, other chiefs among Lamas and other ethnic groups were installed. Tributes collected by the chief of the Lissou ethnic group, for example, were passed to the chief of the Ditchi, who in his turn passed it to the prefect of this king of Lijiang. This network was also availed of by Tais, who lived in territories that spread from the "Canton river to Assam and stretching south to the Malay peninsula."[76] It was from Atuntze that d'Orleans headed southwest for Sadiya, across the Salween River, Khampti, and Dzayul along the Dihing River. What is clear from his narrative is that local ethnic groups such as the Mososs, Lissous, Lamas, Loutses, Chinese, and what he termed "hybrids" were loosely tied to the Lijiang king, and their range of activities show that they were open to nurturing a trade network with ethnic groups from surrounding neighborhoods including the upper Brahmaputra regions.[77]

IV

As the imperial ability to dominate the Brahmaputra was compromised and the ecologically embedded tactical engagements of the local people were in full operation, a lively space of engagements emerged. In the Sirajganj nodal point, where Assam tea and Bengal jute transited for the global commodity market, a remarkable cultural and urban development took place with people arriving here from as far away as Britain and Baghdad, including traders as well as Muslim spiritual figures. Despite the lateral movement of the river that eroded its bank, the rhythm of the flow of people and products remained intact on land as it was on the boats where most transactions took place, making it resemble a "town without houses." A display of a panorama of multiethnic entanglements followed: "The bright head-dresses of the Marwaris (from Rajasthan) afford a lively contrast to the white robes of the Bengalis and the riding costumes and pith hats of the Europeans."[78] Between Bengal and Assam on this waterscape, the contact zones, as Swarupa Gupta has noted, were liminal and amorphous. Gupta traces the writings of a Bengali author, Ramkumar Bidyaratna, who identified himself as "a part of an idyllic realm on the borders of Bengal and Assam, where the waves of the Brahmaputra lured him to a magical Assam wreathed in hope, while the fascination of his motherland Bengal, still lingered."[79] Likewise, Assamese author Lakshmīnātha Bejabaruwā found his life journey inexorably connected to the Brahmaputra's own that spanned Lohit in upper Assam and the Indian Ocean.[80] Bengal's connections with Assam were also cemented by the traders from Chittagong. *Kapurai*, the oil merchants of Assam, were Muslims from Chittagong. Golam Haider established a *tanga* (horse-drawn cart) service between Guwahati and Shillong, followed by the opening of a motor service and oil stores. The company was later sold to his fellow villager from Chittagong, Jomait Ullah, and Sons, which later became a limited company.[81] For the Miya Muslims of lower Bengal, the absorption of Assamese culture, as Sanjib Baruah notes, came through agrarian mobilities along the Brahmaputra basin.[82]

Further north, the colonial administration brought in surveyors, explorers, spies, and military personnel—all triggering flows of different kinds of people around the river network. When Thomas Cooper attempted his trip from Assam to Tibet in 1869, he brought with him four assistants: George Philip, a Chinese Christian; Lowtzang, a Chinese; Masu, a fourteen-year-old Chinese-speaking Tibetan boy who was bought from his mother for 8 taels (6s 8d); and Owhalee, a Chinese-speaking Muslim ex-soldier from Bombay who had served in the Chinese army in Hankow. As Cooper's team arrived in Dibrugarh on the way to Sadiya, he also met a West African man. Originally a sailor, he had joined the Native Artillery Corps in Assam where he married a local woman and became fluent in Assamese. Cooper's racial bias against him was as clear as his appreciation of his "iron constitution and undoubted pluck." He was hired by Cooper as an interpreter as he marched through the Mishmi Hills on the borders of Tibet.[83]

If part of these assemblages of people from disparate places in and beyond Asia were hardly serendipitous, there was also state manipulation bringing people to the Brahmaputra Valley, which shaped part of the social and cultural landscapes. Although susceptible to conditions of hardship and morbidity, workers at the tea and oil fields added more diversity to the ethnic, linguistic, and religious composition.[84] On occasions, such as in a reconnaissance work during World War II, the Rajputs were brought in for excavation work, as the Mishmis were reportedly unavailable to use shovels, picks, and spades. The Mishmis and Abors forged trade relations with other ethnic groups or facilitated their mobility through their network, which also contributed to a cosmopolitan flow. The upper Khampti traders reached up to northern Assam and the Brahmaputra basin, selling ornaments made of silver from mines in Khampti as well as rice and vegetables, making northern Assam "indebted" to them. Even smaller and less well-known groups like the Meyor, living along the watery network of the upper tributaries of the Lohit, developed tripartite trade relations with the Mishmis and the Tibetans.[85] Other transregional trade that ran through Abor or Mishmi networks included the Manangba (Nyishang) from central Nepal, who organized trading trips all the way to Southeast Asia and Hong Kong, particularly from the second half of the nineteenth century and continued through the fateful years of the 1940s. Their routes included Assam (clients included the Nepalese diaspora in Assam tea gardens) and Manipur in northeast India to Burma and Indo-China, bringing in skins, yak tails, manufactured woolen articles, silk mufflers and blankets, musk, and medicinal plants. From Kolkata, they brought needles, safety pins, synthetic dyes, semiprecious stones, and their imitations (from Italy) and sold them in various towns in Assam and Darjeeling, while also smuggling rice into Thailand from Burma.[86]

When the weather improved in the autumn, Zayu County on the Zayu River (which takes the name of Lohit after its two branches converge below Rima) witnessed various ethnic groups assembling from different riverine nodes for work and trade. They included the Darus from the sources of the Irrawaddy, Bebjiyas and Ch from the Dibang, and Mijus (Mishmi) and Taroans from the Lohit. Observing this myriad movement of people across challenging landscapes, Cooper remarked, "It is surprising what physical difficulties trade will sometimes overcome."[87] The annual fair in Sadiya during the first full moon in early February was a rendezvous for various ethnic families from upper areas, drawing more than three thousand people from among the Mishmi, Khampti, Miri, and Singpho (Kachin) men, women, and children. At this meeting point of the valley and the mountains, the visitors not only traded but also enjoyed field sports and racing.[88] Making pilgrimages to the Brahmakunda on the Lohit formed another nodal point of assembly of people from different parts of India and beyond. Besides attending fairs or making pilgrimages, there were people from the plains of India who fled prosecution and found shelter in the regions beyond the Brahmaputra headwaters. Kaulback had a glimpse of this as he camped near a monastery at Singke Gompa, north of the Dibang and west of the Zayul. At the end of their meal, Kaulback was served vegetables and popcorn, which he "had always thought to be exclusively American." In addition to the presence of the Abor and Mishmis, Kaulback met the Babas, the Khampas from the Salween Valley, Popas from central Tibet, and a Bengali girl, whose presence was a mystery to him.[89] A profusion of ethnic mingling, combined with the diverse array of human activities and commodities, was reflected in the use of numerous languages. Colonial official Lambert noted the presence of ten interpreters among them, collectively speaking fifteen different languages.[90]

In such a natural gravity and cultural plurality created by the Brahmaputra network, even the "imperial man on the spot" himself was a willing participant in the flow. In one such example, while carrying out a geodetic survey between the territories around the Brahmaputra and the Chindwin, Irrawaddy's largest tributary, Lambert traveled along the Taiyong River and camped at its confluence with the Taikham River. The luxuriant landscape and provisions, including abundant *mahseer* and *bokha* fishes in the rivers, and the serenity of the wildlife (see, for example, Figure 1.5) made

FIGURE 1.5: A lone deer in the company of birds and a crocodile on a Brahmaputra tributary near Sadiya. Source: T. T. Coopers, *The Mishmee Hills* (1873), 160.

Lambert and his colleagues feel like resigning and camping there for the rest of the year, and he was even poignant enough to pray: "May they [the local people] long be spared from the terrible consequences of Western Civilization."[91] In that phenomenal moment of entanglement with the riverine landscape, Lambert dwelt on a transcending commons that defied a singular power relation or a fixed identity. The cosmopolitan space created at the headwaters of the Brahmaputra was comparable to those in the nodal points at the lower end of the river and the rim of the Bay of Bengal.

The imperial idea of connecting India and China through Sadiya and Lijiang on the Brahmaputra and Yangzi never materialized. But that made little difference to the connectivity that pre-dated and survived the British Empire. Kolkata, Chittagong, and Dhaka in the lower Delta; Sirajganj, Guwahati and Dibrugarh in the midranges of the river; and Sadiya, Rima, Zayul, and Batang at the upper Brahmaputra and beyond—all contributed to an interactive line of communication. Although the Brahmaputra ceased to be navigable just above Sadiya, it continued to inform mobility all the way to Lijiang by the Yangzi to the north, the upper Salween Valley, and the Irrawaddy headwaters via Khampti. These connections were made possible not only by the mobility of ethnic networks but also by their determination to carve out a commercial space amid imperial expansion. The consequent encounters between the British and the local ethnic groups, especially the Abors and the Mishmis, created a common space for mutual, if often uncomfortable, engagements. Colonial flows stumbled at the river's edges, and the local inhabitants had a way to pick up and sustain the flows. As Jelle Wouters argues, the plains were integral to the hill people, not an essential other—an assessment that is echoed in Karlsson's emphasis that these people had the "sensibilities, attachments, and aspirations that cannot be reduced to a negation or the single logic of grid avoidance."[92] The grid-breaking practices among the ethnic groups were informed by their conscious and active efforts in connecting to the flows that passed their territories. The Brahmaputra was a key to such flows that spanned the Bay of Bengal, Tibet, Myanmar, and Yunnan, either through its navigable parts or its extended footprints on the land. Such mobility and interactions were fittingly replicated within the Irrawaddy network, to which we now turn.

TWO

LIKE AN ELEPHANT
The Many Meanings of the Irrawaddy Corridor

SAILING DOWN THE IRRAWADDY in the winter of 1899, Henry Cadell, a seasoned Scottish geographer, compared the river with the Mississippi, Nile, Colorado, Volga, and Ganga in one go.[1] Without undermining the beauty, vivacity, and resources of these rivers, all of which he knew intimately, Cadell put the Irrawaddy on the top of the list, as he felt it had a "little, if not very much, of each of the good qualities of a great river." Cadell believed that the Irrawaddy offered something tasteful—beyond one's dinner—for an architect or archeologist, a biologist or ethnologist, a geologist or mineralogist, an artist or an economist, and finally, for a "large-minded empire-builder who looks to the future."[2] With this profound compliment to the multifaceted utility of the Irrawaddy, poured from the comfortable deck of a steamer, Cadell foreshadowed Myanmar's future nation-space.

The Irrawaddy is both an embodiment and a metaphor for the Myanmar nation. Unlike the Brahmaputra or the Mekong, the main body of the Irrawaddy was undivided, not being sliced into different postcolonial state spaces. This means the river, occupying more than 55 percent of Myanmar's water basin, has an easy sail across the country's national imagination. In the country's transregional aspirations across India and China, the river was not considered the strongest link; it was regarded as a river of "Burma proper," rather than a connecting thread to these neighboring countries. The portrayal of the Irrawaddy as the nation's exclusive waterspace often obscures how the riverscapes attracted transregional forces that linked northeastern India, Yunnan, and parts of mainland Southeast Asia. Amid the economic and political resurgence of India and China, recent literature seeks to identify the "missing pieces" of Myanmar in the spatial puzzle of connectivity in these regions.³ This chapter revisits the studies of the Irrawaddy as a national space and explores it as a site of intense mobility that speaks to the often-forgotten threads of transregional connections.

I

Over a distance of just over two thousand kilometers, from the Andaman Sea to its sources near 28°N, the Irrawaddy River is shaped by several major tributaries (Figure 2.1). Similar to the confluence of the Ganga and Brahmaputra at Goalundo in Bengal, the Chindwin River joins the Irrawaddy on its right bank at 21°31′N, significantly increasing its water volume. Chindwin flowed in the north-south direction at the foothills of the Patkai range, spanning the territories of the Indian states of Manipur, Nagaland, and mid-western Burma in concert with many of the tributaries of the Brahmaputra that flowed in a maze in the Manipur plains.⁴ On its left bank, its tributaries connected the Irrawaddy to Yunnan. Shweli connects to Yunnan's commercial town of Tengchong (formerly Momien). Although the Taping joins the Irrawaddy at Bhamo north of the Irrawaddy-Shweli confluence, it also reaches Tengchong through its subtributary, Takaw, which, in its turn flows close to the Salween river.⁵

Further upstream about 30 kilometers north of Myitkyina, the Irrawaddy is formed by the confluence of two rivers, the Mali and Nmai (Figure 2.3). The Mali flows from a point in the northwest that is close to the Brahmaputra network in northeastern Assam. Namlang, one of the snow-fed and fordable tributaries of the Lohit, with "yelling torrents" in the summer, emerged from the Mali basin and three routes between Khampti and Assam ran along this river.⁶ The easternmost terminus of the Dihing tributary of the Brahmaputra was only a pass away, at Chaukan, from the Mali river, and was reachable following a musk deer's track in the meadow of trees and ground orchids⁷ (Figure 2.2). Mali's proximity to the Brahmaputra is also reflected in the fact that half a dozen "big rivers," according to Kingdon Ward, flowed from the snowy Lohit divide, on the other side of which was the Lohit river, Mishmi Hills, and Tibet. On its way to the confluence with the Irrawaddy, Mali received at least 16 *rame* and 17 *zup*, in addition to numerous tiny streams, while Mali itself, discharging more than 11,000 *cusecs* of water, was navigable "for shallow draught country boats as far north as 'Nsop-zup, and for Kachin rafts a good deal farther."⁸ So much so that Karl Andree, a German geographer, introduced

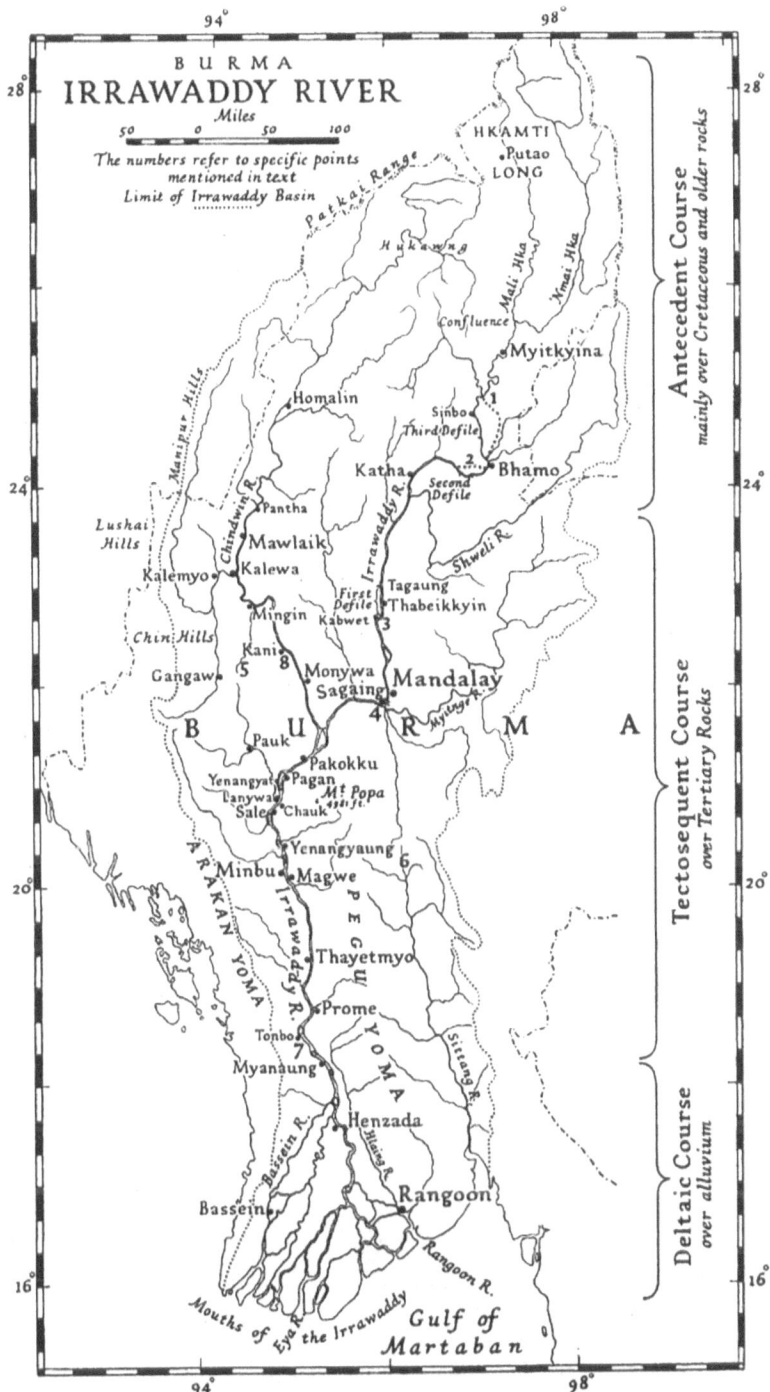

FIGURE 2.1: Map of the Irrawaddy River. Source: L. Dudley Stamp, "The Irrawaddy River," *The Geographical Journal*, 95, no. 5 (May, 1940): 2.

the Irrawaddy's headwaters as being located east of Sadiya, "the point in Assam where Brahmaputra is approachable by steamers."⁹ The long-lasting nineteenth-century idea that the Brahmaputra and the Irrawaddy were the same river was not surprising, as streams flowed in myriad directions between the upper edges of the two rivers.

If the Mali Hka had a penchant for Assam, the Nmai Hka opted for Yunnan. It received at least 8 *rame* and 16 *zup* from the east as it rushed to its confluence with Irrawaddy from the northeast. Some of the *rame* were "big rivers" that "roar down" to the Nmai from the Chinese frontier.¹⁰ Nmai itself had a large tributary, Ngawchang Hka, coming from the Irrawaddy-Salween divide. Fifteen miles further up was the Chipwi River, which reached up to the Panwa Pass into China.¹¹ Shakespear identified the widely used route between Kampti Long and Yunnan across the Mali and Nmai Rivers. A deeper

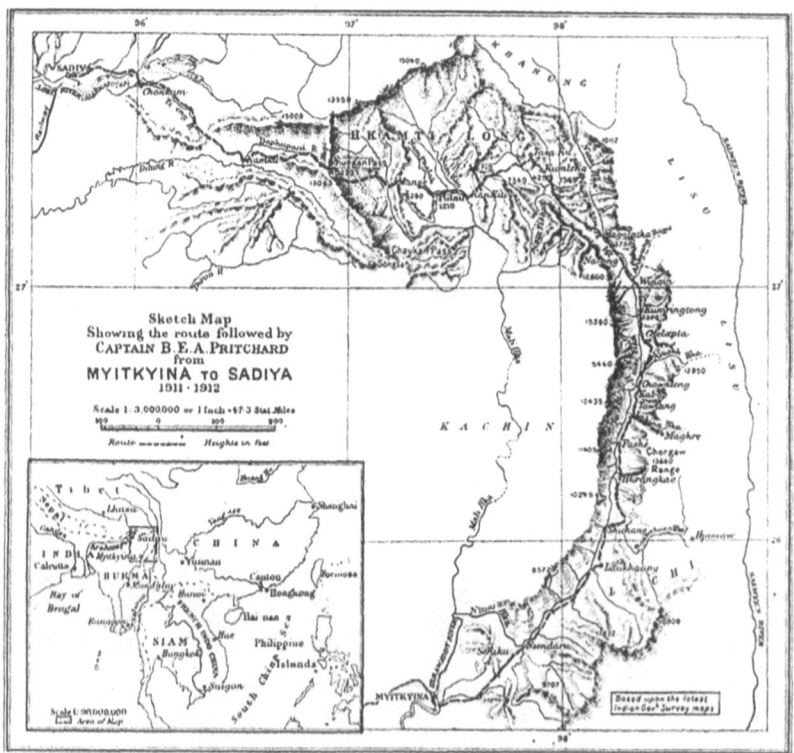

FIGURE 2.2: Routes from Myitkyina to Sadiya. Source: B. E. A. Pritchard, "A Journey from Myitkyina to Sadiya viâ the N'mai Hka and Hkamti Long," *The Geographical Journal*, 43, no. 5 (May 1914).

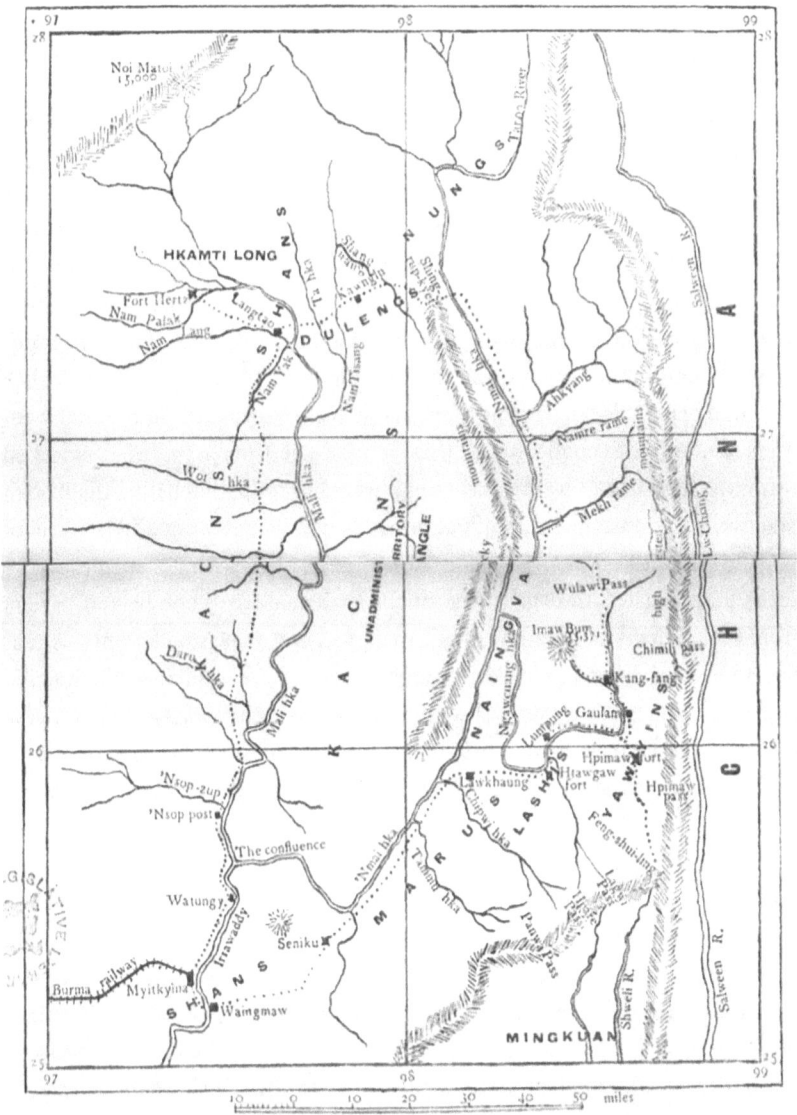

FIGURE 2.3: Upper Irrawaddy tributaries and the confluence that formed the Irrawaddy. Source: F. Kingdon Ward, *In Farthest Burma* (Seeley, Services & Company, 1921), 14.

understanding of the river network would require the exploration of the numerous tiny flows that joined the Nmai, Shweli, and other tributaries of the Irrawaddy that finally made up its main body. "Close around us on every side rose densely wooded mountains which poured ten thousand tributary rills down into the bamboo-choked streams," Shakespear observed.[12]

Within this dense fluvial network, the Irrawaddy announced itself through several nodal points where trade and imperial encounters took shape. Following the annexation of the lower delta region, production and trade in rice boomed to the extent that it became the largest surplus producer by the turn of the twentieth century—a buoyancy that was supplemented by timber trade. A thickening flow of capital and labor marked the new phase of the lower Irrawaddy's economic dynamism that continued until the global economic depression, and the Japanese invasion that followed.[13] Up the Irrawaddy Valley in Yenangyaung, located below Mandalay, another key nodal point rapidly developed after the Second Anglo-Burmese War. Oil fields in this area, enriched by sandstones of the upper Miocene age, were in production before the British takeover and continued to produce about 45,000 barrels annually (1 barrel of 35 gallons), until output drastically rose after the Third Anglo-Burmese War. From 62,721 barrels in 1888, the production reached 6,020,000 barrels in 1898 from the oil fields of Yenangyaung, Yenangyat, and a few smaller ones. About 600 wells, largely drilled by American engineers, were located within an area two miles long and half a mile broad. In the decades previous to the Japanese invasion, these oil fields were producing about 6.5 million barrels annually.[14]

From the perspective of broader transregional appeal, the next significant nodal point was Bhamo. Located about a thousand miles north of Rangoon and at the highest navigable point that stood within 130 miles from Chinese borders, Bhamo (Manmaw in the Shan language, meaning "potters' village") was a hub of Chinese merchants coming overland from Yunnan.

Once the rainy season was over, from November to May, Bhamo saw the arrival of "innumerable strings" of mules, bullocks, and ponies from Yunnan, bringing up the silk and "notions" of China—as John M'Cosh noted. Other than silk and silk products, Chinese products coming to Bhamo included gold and silver bars, brass and copper vessels, mercury, arsenic, vermilion, carpets, fans, spices, rhubarb, musk, and dried fruits. The return cargo included cotton, wool, ivory, edible bird's nests, and British woolen products and calico. John M'Cosh found Bhamo to be a site of a "great stream of Chinese commerce" where caravans of 30 to 40 mules or bullocks constantly arrived and a significant number of Chinese came every year doing business worth £700,000 sterling.[15] From the Burmese side, the flat-bottomed boats from the Irrawaddy downstream brought in cotton, precious stones including amber, and other items. Of about £425,000 sterling of Burmese export, more than 50 percent came from cotton and cotton products.[16] The brisk trade in Bhamo is reflected in the fact that in the spring of 1870, eight hundred loaded mules arrived in Bhamo from Yunnan per month.[17] These statistics were from the years when trade and commerce were in their lowest ebb due to the Hui rebellion in Yunnan against imperial China (1855–1873). The suppression of the rebellion and the British annexation of Burma in 1885 revived the Burma-China trade routes in multiple directions, significantly increasing the value of the trade. In most cases, these figures were understated, as traders usually recorded incorrect returns or evaded registration altogether to avoid paying customs duty.[18]

Myitkyina, located north of Bhamo in a more precarious landscape, was another major nodal point. The confluence of the Mali and Nmai that created the Irrawaddy above Myitkyina was not merely a hydrological affair—it generated the flow of people and commodities to and from India and China, with Myitkyina at the heart of a "helio communication."[19] For example, one of Myitkyina's bazaars, held at Namkham on the Namkhai and Meung Mao route near the border of Yunnan and the northern Shan States, drew people across the frontiers, numbering as many as six thousand.[20] From the vantage of the commercial nodality, Bhamo and Myitkyina were connected to the Irrawaddy as Sirajganj and Sadiya were to the Brahmaputra. Imperial responses to the trading potential of the Irrawaddy and these nodal points along with the connected land routes were also equal to those seen in the Brahmaputra basin. However, these responses manifested in distinct spatial specificities, to which we now turn.

II

With three wars fought and won at regular intervals throughout the nineteenth century—the last of which the British famously described as a "picnic"—the Burmese Konbaung Empire was formally annexed by the British in 1885, marking its dissolution. But the empire building was a staggered process, informed by a combination of interlocking global events, a singular urge to reach out to China, and tough ethnic encounters. This flux of events and agencies found material expression at a few locations within the Irrawaddy system, which shed light on both the advance and limits of the British Empire.

Following the conclusion of the Second Anglo-Burmese War in 1853, the British sway over Burma was maintained by a steamer conglomerate known as the Irrawaddy Flotilla Company (IFC). It came into existence in 1865, partly in response to the growing global prospect of Indian Ocean trade due to the imminent opening of the Suez Canal. The company's directors believed that there was "no trade to the East more capable of, or more likely to have continued expansion, than that with Burmah."[21] With the opening of the Suez Canal, the operational success and profit of the company exceeded "the anticipations of the original promoters."[22] The bulk of trade was influenced not only by enhanced mobility in the Indian Ocean but also by domestic demands from upper Burma, an often-overlooked aspect of the region's maritime history. About one-third of the total rice produced in British Burma found its way to upper Burma, indicating that both the external and internal demands for commodity transportation favored the IFC.

In 1870, the total value of exports of rice, teak, cutch, and cotton from Rangoon on the IFC steamers amounted to £3,603,698 sterling and imports were worth £3,615,554 sterling, while the company profits and fleet size showed steady growth.[23] From a modest beginning with four fragile flats ("shallow draft barges with covered deck") leased from the government, the IFC commissioned 10 steamers and 18 flats by 1872, and on the eve of World War II, the company owned about 600 ships, nearly all of them originating in Glasgow shipyards. If the IFC catered to the demands of the imperial hotspots in Rangoon, Yenangyaung, and Bhamo, among other places on the Irrawaddy, its ultimate goal was to facilitate British-Burmese trade to China. Within three years of its establishment, it started operations from Rangoon to Bhamo, aiming for "re-opening the ancient, and once extensive trade by the Irrawaddy to and from Western China."[24] The annexation of northern Burma in 1885 provided the much-expected boost to such a plan.

Yet, despite encouraging profit margins, the IFC could hardly replace the local transport system on the Irrawaddy. It could not keep pace with the demand for rice transport, which continued to be met by about twenty-five thousand local boats, with the company's fleet size being "minuscule" in comparison. Many boats were sourced at Yandoon, fifty miles up from Rangoon, which was populated by a large number of boat-owning local merchants. Up the river, Yegin, a major rice-producing district with trade links to upper Burma, exported about fifty thousand baskets, but the IFC steamers were unable to establish a foothold there. Further up at Yenangyaung, at the oil fields, a large quantity of oil used to be extracted three miles off the river bank and was mainly carried by local boats. At least a third of the rice exported from Rangoon went to upper Burma on "immense fleets of boats," as the flats of the IFC could not compete with the lower freight charge for local boats. Likewise, it could only transport half of the salt traded.[25] At Minhala and Magway (where peas, grain, wheat, cutch, cotton, and teel-seed oil were produced), no trade was secured by the IFC either. Even regular passenger and commodity transportation costs were higher than boat transportation and, on most occasions, even the Burmese elites preferred transportation via boat and cart to railways.[26]

In the relatively wider creeks where steamers could make headway, the challenges of sandbanks, no-tide-no-navigation situation, sharp bends, shallowness and overgrown weeds and tree branches stalled the passing of steamers beyond the mainstream of the Irrawaddy. In the lower Irrawaddy, of the 26 "Creek Steamers," 19 suffered from these disadvantages.[27] The Bassein Creek, which was larger than other creeks, was the main waterway between the Rangoon and China Bakir Rivers, but steamers could only enter the creek for less than five hours a day depending on the tide, owing to the shallow waters causing delayed navigation and higher cost of transportation.[28] In 1915, about five decades after the introduction of the IFC, the Irrawaddy Delta districts were annually producing 2,143,000 tons of unhusked rice collected from around 4,076 square miles of areas in Pegu, Insein, and Tharrawaddy, but the IFC dealt with only 240,000 tons of paddy. The railways carried a similar amount; the remaining 1,743,000 tons were transported on local boats frequenting all tidal creeks, each with a capacity of 1,800 to 2,500 baskets. Thus, what emerged as a flagship navigational enterprise of the British Empire soon found itself inadequate to compete with the indigenous river crafts in handling commercial demands. Finally, the Irrawaddy that launched the career of the IFC was also where it was sunk. Due to the fear of the Japanese invading upper Burma via the Irrawaddy, most of the 600 ships, flats, and steamers were either sunk or scuttled. This action mirrored what the British administration did to more than 25,000 boats in the Bengal Delta due to a similar fear of a Japanese invasion, which partly contributed to the catastrophic famine of 1943.[29]

If the British had a hard time controlling the commercial flows and routes on the navigable Irrawaddy, their imperial might was truly tested in the northern parts of the Irrawaddy. The Chinese had been watching the developments in the Irrawaddy corridor since the First Anglo-Burmese War and were already apprehensive about British intentions. Captain Hannay's exploratory trip to the upper Irrawaddy in 1836 drew a strong reaction from the Qing imperial court and Yunnan's provincial administration, which compared the English to a "pipul tree" (which spread fast and is difficult to eradicate) and braced themselves for resisting them in the region.[30] The apprehensions were particularly fueled by the fear of losing influence in Bhamo, being an ancient center for Chinese trade that the Chinese described as Jiangtou Cheng (Riverhead Town).[31] Hannay's description of it reflected a sort of exclusive economic zone for China:

> No foreigners except the Chinese are allowed to navigate the Irrawaddy above the Choki of Tsampaynago, situated about seventy miles above Ava, and no native of the country even is permitted to proceed above that post, except under a special license from the [Burmese] Government. The trade to the north of Ava is entirely in the hands of the Chinese, and the individuals of that nation residing at Ava have always been vigilant in trying to prevent any interference with their monopoly.[32]

The closing-in of the British power at the Chinese frontier following the Third Anglo-Burmese War, therefore, triggered complex and lengthy negotiations on border demarcations in the region adjacent to Bhamo. Because its eastern coast was increasingly becoming inaccessible via Shanghai and Canton, China sought to access the Indian Ocean via Rangoon through tense negotiations with the British that centered on the question of Qing China's access to the Irrawaddy system. In the 1892 border meetings, China demanded access to the Irrawaddy by Chinese ships via Bhamo, in addition to asking for ownership of the territory between the Taping River and Mole, as a "reward" for its cooperation with the English during the annexation of upper Burma. They also insisted on the right to free navigation from Bhamo for oceangoing vessels bearing the Chinese flag in the manner that foreign ships came up the Thames below London Bridge.[33] This was consistent with their demand to push the boundary to the west and southwest as close as possible to the Irrawaddy Valley to retain control of most of the Taping and the Shweli Rivers. They asserted that any other British concessions against Chinese territorial claims, such as Kiang Hung in the Mekong basin, would not be equivalent to the withdrawal of its demand for the territory east of the Irrawaddy.[34]

Despite the protracted negotiations, very little precise demarcation of borders could be made, not merely because both sides disagreed on many policy issues, but also because tracing the tributaries of the Irrawaddy and the Salween was beyond contemporary cartographic skills. In the Pang Long area in the southern Shan States, all of the eight cairns fell on some form of running bodies of water, either at the confluence of two streams or at the crossing of a river and road or at the ridge of a headwater—where any claim to sovereignty was as fluid as the bodies of water themselves.[35] The Irrawaddy's navigable parts remained a site of the British imperial network, but the dream of using it as a springboard for free rides into China was frustrated not only by China's pressure to secure the headwaters of the Irrawaddy bordering Yunnan but also by its demand for access to the entire course of the river. On the other hand, at a wider gaze at Asian water spaces, the British entertained the possibility of connecting the upper Salween, Mekong, and Yangzi Valley via the Irrawaddy. This Anglo-Chinese imperial vision of accessing multiple riverscapes survived two world wars and Chinese nationalist rule. Right after World War II, for example, despite securing the eastern China coast from the Japanese, China proposed to the Allied Forces to establish their consulates at Margherita in Assam on the Brahmaputra and Myitkyina on the Irrawaddy, which the Allied Forces officials thought "may not be wholly without significance."[36]

While the imperial forces kept negotiating their influence over the Irrawaddy in the changing geopolitical circumstances over more than a century—which resulted in operations of some sort of "shadow states"—a third force that inhabited those water spaces made themselves clearly visible.[37] For the British and the Chinese, the trade routes between the navigable points of the Irrawaddy, Salween, and Mekong Valleys on the same latitude ran through the territories of a range of local ethnic groups who self-consciously asserted their rights. Despite the euphoria about imperial profits around Suez and Panama Canals, the fourfold increase of Burma's trade happened with ten Asian territories rather than with the rest of the world, India being the leading one.[38] These flows were reflected in the coastal trade around the Bay of Bengal and North China Sea, but in a larger picture these were a reflection of trade that was carried out in the river routes and the connected landed nodal points across the river valleys. An explanation of these remarkable flows requires us to study the local agencies that operated inland between the frontiers of Burma and China with the Irrawaddy system as a key catalyst.

Like the Abors and Mishmis on the Brahmaputra headwaters, there were groups who guarded the headwaters of the Irrawaddy and parts of the Salween and connected land routes. An exploration of encounters between the local forces and the imperial advances is important to understand the creative entanglements of the local and the global. As the British and the Chinese initiated border talks soon after the British annexation of upper Burma, MacMahon, an English commentator, noted a "more than ordinary interest" attached to these ethnic groups simply because they had dominated the trade routes between Bhamo and Tengchong in Yunnan. He lamented that the idea of "surrendering an important entrepot of trade and a strategical position like Bhamo, which not only controls the whole of the Upper Irawadi valley, but is also the natural centre for railway and telegraph lines between India, Burma, and China, seems absolutely preposterous."[39] Yet this uncomfortable adjustment was what defined British imperial engagements along the routes that connected Yunnan with the Irrawaddy. The Chinese were equally on slippery ground as far as these ethnic groups were concerned. What follows is an exploration of the high powers' encounters with these intervalley ethnic groups with a focus on the Kachins (originally Kakhyen, meaning the "People of the Headwaters") who epitomized the complexity of this relationship.

III

With the conclusion of the Second Anglo-Burmese War, the British and the Chinse imperial frontiers came in even closer range, but the ethnic groups that stood in between loomed large. For Edward Sladen, the leader of the first major English commercial reconnaissance mission, it was important to identify the "exact position held by the Kakhyens, Shans, and Panthays, with reference to that traffic, and their disposition, or otherwise, to resuscitate it."[40] Despite warnings from the Burmese officials about a possible ambush by the Kachins, Sladen had a different experience with them. He observed that their hostility toward the travelers on trade routes passing through their territories was a bargaining chip to regain access to Bhamo, from which they had been temporarily barred by the Burmese king. Although Sladen showed clear racial prejudices against the Kachins, he specifically noted their passion for trade and hospitality to strangers. He also mentioned the cordial commercial relationship of the Kachins with their Shan neighbors, as seen in the towns and marketplaces across the northern Shan States.[41] On his way back from Yunnan, when he reached the confluence of the Taping and the Irrawaddy, Sladen took a parting glance at the Kachin Hills and noted that the two peaks, Shitee-doung and Kad-doung, standing on both sides of the Taping River were like sentinels guarding the routes to China.[42]

At least a million-strong in the 1910s, the Kachins literally guarded the routes that connected Bhamo with Tengchong. They maintained a strong presence from the Mali-Nmai confluence above Myitkyina through all the territories in the "triangle" (Figure 2.3). The Kachins, who inhabited this triangle, were connected to the fluvial network of the Irrawaddy, Brahmaputra, and Salween, and enjoyed a higher status than the Kachins who lived outside the triangle.[43] A 1916 survey found that both Kachins and Shan headmen managed twenty-two ferries along the 180-mile stretch of both the banks of Mali that spanned clusters of villages. The importance of each of the ferries was designated according to the capabilities of transporting mules and the commodities attached to them. A small amount of toll was collected by village headmen, sometimes in cash and sometimes in kind for salt and rice, and sometimes no toll was collected at all. All ferry terminals were located near the confluence of smaller tributaries with the Mali River.[44]

The advantages that the strategic locations offered to the Kachins turned them into conscious actors, who sought to promote their political careers within the community by way of collecting "tributes" or "tolls" from traders and lowlanders in exchange for protecting their caravans between Burma and Yunnan. Because of this, the Kachins have often been labeled as a "protection racket," as James Scott and David Leach have observed.[45] Yet it is also possible to read the Kachin internal political practices as an evolving outcome of a more ambitious encounter with the external world of transregional trade that the Burma-Yunnan corridors offered them. The Kachins, like the Abors and Mishmis on the Brahmaputra, sustained a parallel network to entangle themselves in the flows of trade. Beyond collecting tolls, a Kachin *sawbwa* (chief) was usually a trader who profited from the hiring of mules or coolies for transport.[46] The fact that the Kachin local chiefs (*duwas*) owned the jade mines in the Mali-Nmai areas and offered excavation rights to about thirty thousand Yunnanese Han miners meant that they had vocations beyond receiving indiscriminate tributes. The Kachin's rights to the region's jade mines were never questioned by the British-Burmese or Chinese governments—which also indicated that the Kachins had legitimate stakes in these territories.[47]

Despite occasional conflicts, the Burmese king maintained close relations with the Kachins to keep their trade routes to China open. Various facilities, including rest houses, were provided by the Burmese authorities to those traveling to Bhamo. Sladen saw the wisdom of this policy and suggested that both the Chinese and the Burmese should, in their respective borders, offer to the

Kachins "a proper sum, in recognition of their territorial dues" and for safe passage through the trade route; Sladen was also eager to "cultivate an entente cordiale" with them.[48] After the annexation, the British administration followed through with this policy of paying them a fixed toll.[49]

The diplomatic conflicts between British Burma and China over the borders made little impact on the political and social space of those living on the borderlines, as British officials would often abandon "disputes over minute details." On one occasion, for example, the Burma-China Boundary Commission accepted a situation in which Shan and Kachin villages came within the agreed British border but were allowed to enjoy the "graceful concessions" of retaining their paddy lands, created out of the shifting bed of the Nam Yang river within the Chinese border.[50] Concerning settling the Burma-China frontier, the lieutenant governor to Burma proposed to the viceroy to India that he did not find it worthwhile to spend men and money to serve a specific British interest, as he felt it was of "no consequence whether the status quo is allowed to continue till an agreement is arranged at Peking. We shall be no worse off and locally our prestige will not suffer."[51] The main beneficiaries of this ambivalent relationship with China were the Kachins, among other ethnic groups of these borderland areas.

China itself was no less complacent in this context. When imperial China was at war with the Hui rebels in Yunnan, the Kachins sided with the Hui and concluded a formal agreement to "protect and afford safe-conduct to traders and travellers who might have occasion to cross the hill ranges between Bhamo and the Northern Shan States."[52] In exchange for the withdrawal of the British troops from Sadon, the Chinese government assured the British that it would keep the tribes on the east of the Irrawaddy and north of the Burmese frontier under "effective control, and will be responsible for preventing any raids into the Burmese territory or other annoyances." Yet they acknowledged that they rarely exercised effective control over the ethnic groups. The Chinese representative to the negotiations, Macartney, informed the British-Burmese negotiators that Chinese authority had been sporadic, stating that China "only interfered with the tribes when necessary."[53] Even in serious conflicts that arose with the Kachins on opium-related issues in 1939, China was unable to defeat them.[54] Soon thereafter, during the full-fledged war with the Japanese, the Anglo-Chinese alliance found Kachins, along with the Shans, "absolutely necessary" for carrying on military communications between Bhamo and Yunnan, as they alone were believed capable of surviving the deadly malaria epidemics, unlike the Chinese who were vulnerable to this disease.[55]

The numerous Kachin subgroups were clustered as Kakoo-Kanams, or upper and lower Kachins, referring to their respective positions along the upper Irrawaddy. The dividing line was generally the third defile of the Irrawaddy, but in the mid-nineteenth century, the Kachins were pushing further south along the river, below the second defile[56] (Figure 2.1). However, the Kachin network was not confined to the Irrawaddy headwaters and the Yunnan borders. On the western side, from the Brahmaputra headwaters to the Chindwin Valley, they were equally visible and vigorous. They inhabited both sides of the Patkoi range, their older home being the Hukong Valley, fed by the upper Chindwin River, while also being spread across Khampti in the north, the Naga Hills and Sadiya on the west, and Burma proper in the south.[57] In other words, the Jinghpaws (Singpho in Assam, meaning "man par excellence"), the largest Kachin clan, were well placed to connect with both the upper Brahmaputra and upper Irrawaddy tributaries.[58] Despite their strategic decision to side with the defeated Burmese forces against the British in the First Anglo-Burmese War, they remained in control of the trade routes between the upper valley of the Brahmaputra and the Irrawaddy, as well as the western bank of the Salween.[59]

The Kachins were aware of the commercial prospect of the entire Irrawaddy Valley and were often as ambitious as a state in their contemplation of the long-term trade flows. In the course of the nineteenth century they moved south from the headwaters, and, had the British not headed north, they would have reached Mandalay in light of the increasing weakness of the Burmese Konbaung dynasty. And in that sense, they were more reticulate than some anthropologists, including Leach, have noted, for there was a "Kachin perspective centred not only upon the Shan but also upon, for instance, the major Burmese throne itself."[60] For the Kachins, therefore, there were broader issues to deal with than merely collecting tolls from caravans passing through their neighborhood, which included negotiating access to mineral resources by outsiders, strategic encounters with imperial powers, and sustaining social fabrics that were informed more by the externalities of engagements than by domestic fixations. The Irrawaddy system proved to be the major impetus for these larger spatial and material protocols of the Kachins.

Like the Abors and the Mishmis on the Brahmaputra headwaters, the Kachins asserted themselves as astute guardians of the Irrawaddy headwaters and neighboring areas where principal trade routes between Bhamo and Yunnan ran. Recent studies have focused on the networks that the Chinese commercial groups developed between China and mainland Southeast Asia, appearing as an "arc."[61] The British imperial transport network, exemplified by the IFC, significantly enhanced mobility. This is evident from the government mandate requiring the IFC steamers to carry at least fifty deck passengers (emigrants) free of cost.[62] So it seems that the British and the Chinese empires, with all their strengths and weaknesses, developed and deployed a network that was alive only with the active participation of the local inhabitants living along the upper Irrawaddy and neighboring basins.

IV

Within the common space that emerged through the entanglements of European, local, and regional actors, new forms of cosmopolitanism took shape, which was also aided by a new wave of economic activities in the Irrawaddy delta. A whirlpool of activities by Burmese cultivators working the marshy fields, Burmese girls stitching rice bags of Bengal jute, Madrasi coolies donning the loaded rice bags for the ships, Scottish engineers preparing the ships for departure, Chittagonian firemen and *khalasis* and Chinese carpenters busy at their work—all added to a cosmopolitan flow that was as messy as lively. But the rice was not the only prompter of such remarkable human mobilities. Traders including "native-born, half-caste, Parses, Armenians, or Chittagonians" and others from the ports of Calcutta, Bombay, Madras, Moulmein, Singapore, and Hong Kong continued to throng in Martaban, Rangoon, and Prome to trade in such diverse items as raw cotton, petroleum, cutch, lacquered ware, bell-metal, and rubies in exchange for British or British-Indian products.[63] By the turn of the twentieth century, Rangoon had become the third-busiest port after Bombay and Calcutta. This economic buoyancy attracted a vibrant mix of intellectual, cultural, and political actors whose activities flourished until the Japanese occupation, a glimpse of which is captured in Figure 2.4.[64]

Upstream, Mandalay had a character of its own, representing the continuity of precolonial cultural interactions with China and India, much of which is reflected in the diary of Wekmasuk Wundauk U Latt, a late nineteenth-century Burmese elite and other local authors. Mandalay was more of a cultural hub than a commercial one. With the advancing British posing a challenge, there was an urgent need to preserve the cosmopolitan ethos that had been ingrained in its culture since the last days of the Konbaug dynasty. In the new circumstances, the example of U Latt, who bought a jasmine garland from

FIGURE 2.4: Steamers and native boats on the Irrawaddy, at Thayatmyo, Burma, 1907. Source: Library of Congress: https://www.loc.gov/resource/stereo.1s17693/.

a Brahman to offer it to the Buddha statue in his house, was complemented by the example of the king's tolerance of Burmese converts to Christianity.[65]

Further up in Bhamo and beyond, diversity and plurality of cultural engagements hardly waned. Yule described Bhamo as an "emporium thronged with cotton bales and bundles of silk, with pale Chinamen, black-jacketed Shans, and all the trafficking tribes of those obscure regions."[66] Australian traveler George Morrison noted a "wonderful mixture of types" and quipped that "nowhere in the world, not even in Macao, is there a greater intermingling of races. Here live in cheerful promiscuity Britishers and Chinese, Shans and Kachins, Sikhs and Madrasis, Punjabis, Arabs, German Jews and French adventurers, American missionaries and Japanese ladies."[67] Bhamo enjoyed a "double life," noted O'Connor, a Bengal-born Irish traveler, because it was quieter in the rainy season and sprang to life afterward, inviting tides of humans, mules, and commodities. O'Connor observed hundreds of caravanners and thousands of mules, along with various merchants such as an Indian Muslim named Sheikh Ibrahim, a Chinese named Ah Tan, and a Jew running a billiard saloon. Yunnanese merchants were a major force in this flux and had deeper entrenchment than seasonal travelers. O'Connor met three brothers from Tengchong, who took turns spending three years each in Bhamo. They maintained homes in Tengchong, Bhamo, and Rangoon.[68]

Bhamo's cosmopolitan vibe continued to travel to Yunnan via Tengchong along both the Taping and Shweli Rivers, whose banks were dotted with Kachin or Shan caravanserai, creating spaces for "microcosmopolis" along the river paths. "Pathways lead to it through the heart of the river-jungle, where the purple Taping, laden with the waters of Momein, steals through waving grasses to its union with the Irrawaddy," O'Connor noted.[69] On the Shweli Valley, not far from Bhamo, a "great fair" was held annually in Namkwam. Shakespear discovered a large market where

> lines and lines of booths were crowded with thousands of wild, strange types of humanity—Burmese, Chinese, Shans, Kachins, and Yawyins, their women . . . silk, weapons, homemade cloths beloved and distinctive of the different tribes . . . tawdry Birmingham and Manchester goods and cheap American cigarettes in thousands! . . . Confectioners, sports presided over by Chinese doing a roaring trade . . . nearby going on cattle fair with large numbers of excellent little Shan ponies, mules, and cows picketed in long lines for sale.[70]

As one traces the Taping and Shweli rivers to their sources, they encounter the vast network of the next major river, Salween, along with the continuous flow of people and products from further east crossing the Salween and beyond. At the village of Mang-yam below the confluence of the Che-wen stream with the Laking Hka, Kingdon Ward met five Chinese peddlers from the Mekong Valley who had crossed over the Mekong and the Salween to reach the Lakhe Pass and the Laking Valley to sell salt, cloth, and iron cooking pots.[71] On his way back to Burma along the Shweli River Valley, Ward met with another Chinese woman selling earrings that she had collected at Lijiang, famous for its extravagant fair at the temple of the Water Dragon.[72] On another trip a few years later, in May 1926, Ward traveled from the Shan plains of the Khampti to the last Kachin Hills and the confluence of Seinghku and Adung Rivers that formed the Tamai River. Here he met two Chinese peddlers along with their coolies, selling salt and cloths coming from Weihsi on the Mekong, who traveled along "all these low-lying valleys at the sources of the Irrawaddy" right through the winter. Some spent several years in the region, but most of them visited annually, reinvesting in the produce of the region, including skins of megabats (*Pteropus*) and the root of *Coptis teeta* (*hwang lien*), a useful Chinese febrifuge, and returned to China "with loads as heavy as those which they brought."[73]

The mobility of people and commodities along the headwaters of the Irrawaddy to the Salween and the Mekong Valley helps demystify the process of ethnic identity formation, which was often more than structural. In the upper Irrawaddy region, the line of the water spaces offered a propensity for translocal and transregional connections and opened new windows for layered identities. Anthropologist Pritchard, in his trip from Myitkyina to Sadiya along the Nmai River, was told by the Naingvaws that if they "went further down the river they became Marus, and if further still they became Kachins."[74] Kingdon Ward corroborates by noting that the Marus of the upper Nmai Valley are Naingvaws and that the distinction was geographical, not racial.[75] In contrast to Leach's observations, Daniels and Ma understands ethnicity in western Yunnan as being reconstructed in the context of factors such as long term interaction between communities, administrative change, and "overland trading networks that extended to the Ayeyarwaddy River region in northern Burma."[76]

One can speculate on how the dietary customs, such as those found across the Irrawaddy basin, manifested themselves in disparate places. In the lower Burma delta, "Indo-Chinese races" prepared with devotion the paste of mashed and pickled fish, resembling a sort of shrimp paste, which was their favorite recipe. Yule noted it as the *blachang* (*balachaung*) of the Malays and suggested that the "putrescent fish, in some shape or other, is a characteristic article of diet among all these races from the mountains of Sylhet [northeast Bangladesh] to the isles of the Archipelago."[77] The river corridor was also the corridor of languages, as the river-bound mobilities and entanglements mirrored multitude of linguistic encounters. It was common to hear Tai dialects spoken identically in the Nam-kiu, a western branch of the Irrawaddy in Khampti, as well as on the banks of the Canton, Mekong, and the mouth of the Chao Phraya River in Thailand and parts of central Yunnan.[78] Layered intertranslatability became part of the mundane life of the people on the move, as well as spatially located entanglements, as seen in the way d'Orléans communicated with local people during his travels from Khampti to Assam:

> We spoke in Latin to Joseph; he spoke in Chinese to one of our Tibetans, Siranseli, formerly a gold miner; Siranseli passed on the communication in Lissu to one of the native porters from the mountains; the porter spoke in his own tongue to a Thai who understood it and the last-named repeated the sentence to the Khamti chiefs. These latter seemed to consider their behaviour quite natural, for in the midst of their incessant demands, they smiled most benignly on us, and even indulged in banter.[79]

The transregional mobility and cultural complex that the people on the Irrawaddy basin represented were deeply embedded in the lifeworld of the river itself, and it thus made sense that they would treat the river network as a common corridor. The combination of the fluvial network and local-global interactions was merely a supplement to a larger space of attachments. They were well aware of the locations ideal for grain preparation and the prime spot for fishing, and even employed a "curiously primitive system of insurance against famine."[80] Maru headmen would, therefore, consider it their duty to keep two hundred yards of intervillage roadway up to the Nmai River.[81] Similar perceptions of the water spaces as a commons made the Kachins assiduously defend them from the British government's attempts to survey or obtain a cross-section of the confluence of the Mali and Nmai Rivers that created the Irrawaddy, arguing that such interventions would rouse the ire of the lords of the confluence (*nats*).[82] These practices and the feel of the mystique and animatic reverence for the Irrawaddy were not a mere reflection of the human urge to secure material livelihood and wellness. The riverine space—and the human and nonhuman lifeworlds it invited—held a captivating allure in human consciousness, equally enticing for the *nats*-loving Kachin and the exploratory spirit of a colonizing European (see, for example, Figure 2.5). As Pritchard noted after completing his exploration of the upper Irrawaddy: "The charm of these wild surroundings remains with one always, and will not cease to call one back to the rushing torrents and the lofty mountain ranges which are encountered ere the goal is reached where the Great River of Burma has its birth."[83]

FIGURE 2.5: Everyday life of humans and elephants on the bank of the Irrawaddy. Source: J. W. Palmer, *Up and Down the Irrawaddi or The Golden Dagon: Being Passages of Adventure in the Burman Empire* (Rudd & Carleton, 1859).

The Irrawaddy lived many lives—a life of an expansive navigable body of water that stoked national imagination, but also a life that extended beyond the nation, as it informed transregional crossings outside its navigable arteries. It saw human and commodity flows to and from eastern Tibet, northeast India, and Yunnan—linkages that spanned the upper Brahmaputra and Salween, but also drew flows from as far distant nodal points as the upper reaches of the Mekong, Red, and Yangzi, as will be seen in the following chapters. This convergence of human mobilities and commercial flows around the Irrawaddy meant that despite being under the British imperial shadow for over a century, it was hardly exclusively theirs, nor of the other contending powers like the Chinese or the Japanese, who had a short war-time presence around the river. Amid intense political upheavals, the Irrawaddy basin was an open enough space for lingering connections and longings. The next chapter shows how this interactive, translocal, and transregional lifeworld inhabited the Salween.

THREE

SALWEEN
The Angry River and the Taming of the Empire

THE SALWEEN WAS THE naughtiest of Tibetan rivers, as far as the colonial perception of its navigability was concerned.[1] Of its more than three thousand kilometers of flow between Moulmein (Mawlamyine) at the Gulf of Martaban and its point of origin in Tibet, only about one hundred kilometers down the coast were considered thoroughly navigable. Although some parts of the river were quieter in the Shan States, a reliable thoroughfare was interrupted by the current "as swift as a mill race or a lasher" or "Niagara-like fall." upended by a series of rapids running through deep and narrow valleys.[2] By all means, the Salween did justice to its Chinese name, Nu Jiang, which means "the angry river."

Of the three broad sets of historical queries around the Salween, one identifies the river as a marker of fluid and contested boundaries between British Burma, Yunnan, and Thailand—a true reflection of the unruly nature of the river. A dense network of mountains and streams marked most of its landscape, prompting Hallett, a British railway engineer and author, to quip that the region looked like "a chopping sea of hills, in which it would be impossible, without actual survey, to settle the direction of the drainage."[3] It has been described as the "most uncompromising natural boundary in the world," requiring "every fork" of the river to be surveyed to find a boundary for Thailand and its neighbors, as Thongchai Winichakul noted.[4] Another theme on the Salween revolves around the issue of resource exploitation, ranging from timber extraction in the colonial period to hydrological projects in more recent times.[5] A third area of exploration highlights how the Salween hindered the conceptualization of Myanmar as a unified national space. Acting as a "formidable barrier," the river effectively divided the country into two, obstructing easy communication between the Shans of the Cis-Salween and the Trans-Salween regions.[6]

Reading the Salween as a space that exposed colonial intractability and as a doorstop to territorial nationalism reveals its potential as a facilitator of messy and unbounded mobilities. Its unnavigability exposed the empire's relative immobility further downstream compared to other BISMRY rivers, leading to a wider spatial distribution of both local resistance and collaborative entanglements. If its unnavigability announced a level playing field for political contestations between the empire and the locals, the Salween network also offered transregional pathways for people and commodities across territories now located in Myanmar, Thailand, Laos, Yunnan, and Tibet.

I

Between its northernmost major town of Chamdo in Tibet and its exit to the Gulf of Martaban at Moulmein, the Salween spanned Yunnan, Wa, Shan, Karenni, and Mon territories. While the river flowed past key nodal points such as Taunggyi, Lashio, Kunlon, Baoshan, and Dali, its transregional reach extended beyond these hubs through its numerous tributaries. Rima on the Lohit (a northeast tributary of the Brahmaputra) was less than one hundred kilometers. Further down, the Taping and the Shweli—two of the Irrawaddy's major tributaries—reached out to Tengchong, which was located close to the right bank of the Salween. The Irrawaddy's northeastern subtributary of Myit-Nae, below Mandalay, reached out to the Salween and narrowly missed a direct contact. While on its right bank, the Salween was close to the Brahmaputra, Irrawaddy, and Sittang systems, on its left bank at least six tributaries of the Mekong originated. Another intervalley connection between the Salween, Mekong, and Red systems was via Takao and Simao. A third river network connected the Salween at Kunlon with the Mekong via the Namsung River network (Figure 3.1). Further down, another major network existed between the Salween and the Chao Phraya that was connected with the Mekong basin at Luang Prabang. Although most of the main flows of the Salween were unnavigable, some of its tributaries in northern Shan States were partly navigable, one being for "many day's journey" between Mon and Dahgwinzeik, albeit bearing only local significance.[7] In some cases, "fair-sized" streams joined the Salween, which included the Balu Chaung or Nam Lak, Nam Pawn,

Nam Teng, and Ben Chaung or Nam Pang (which joins the Salween ten miles below Kenghkam, where it was a quarter of a mile broad).[8]

Within this intricate fluvial network of the Salween, Moulmein was a major commercial nodal point. In the mid-1840s, the annual imports and exports at this port were worth 16,00,000 rupees and 9,00,000 rupees respectively, one-fourth of the former representing European goods and one-half of the latter consisting of teak, dealt with by thirteen timber merchants. Beside the export trade of teak, the ship-building industry developed significantly between 1830 and 1843, during which 75 ships were constructed, including 58 vessels with a capacity of over 100 tons. Other economic activities in Moulmein included those of the auctioneers, boot and shoe makers, brass founders, carpenters, cloth merchants, farriers, gold and silversmiths, gun and blacksmiths, joiners, and 74 hackney coaches.[9]

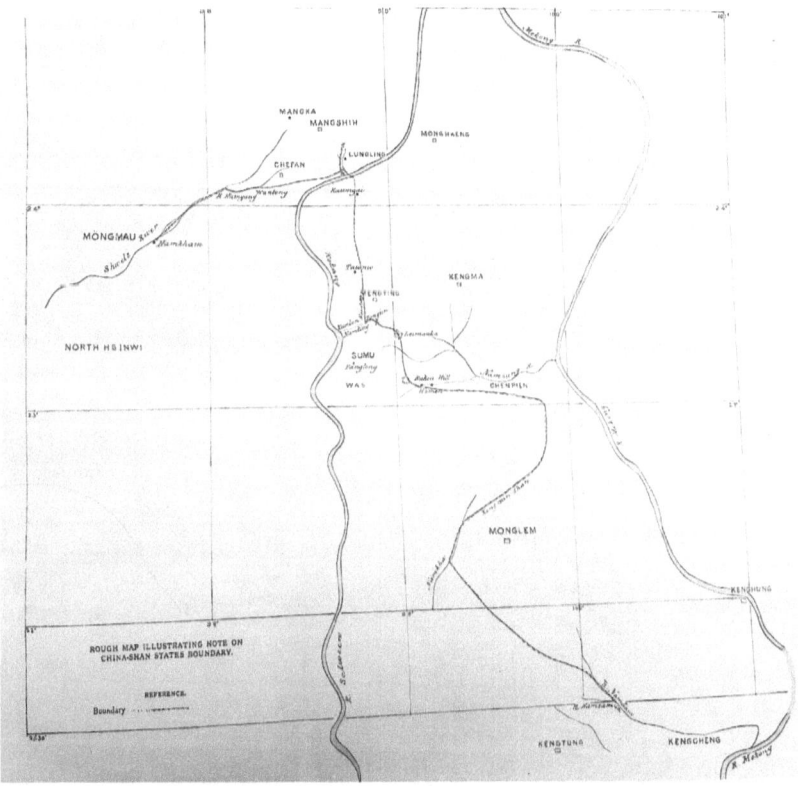

FIGURE 3.1: Upper Salween–upper Mekong connections. Source: London, The National Archives: MP. KK. 49.

Moulmein's connection to Bangkok was mainly secured through the Chao Phraya (Menam or "Mother River" in Thai) system via its northernmost navigable point at Uttaradit in Chiang Mai. The 200-kilometer stretch of land between the two river systems was traversed by ponies, elephants, oxen, and mules. Since the distance between Uttaradit and Bangkok was about 370 miles, this route saved about 3,500 kilometers of oceanic journey compared to the route between Moulmein and Bangkok by the sea circumventing the Isthmus of Kra. A late nineteenth-century traveler saw the "largest river boats" plying from Uttaradit to Bangkok, which reflected the facilities and bulk of trade on this route.[10] As we will see in the next chapter, Uttaradit, being also close to the Mekong network, saw remarkable traffic from the latter (Figures 3.2 and 4.2). This fluvial junction at Uttaradit that connected commercial flows from the Salween and the Mekong was at the heart of social and economic vibrancy of surrounding areas, of which Chiang Mai was most prominent.

By the turn of the twentieth century, Chiang Mai, with a population of about 700,000, experienced a continuous flow of people, products, and pack

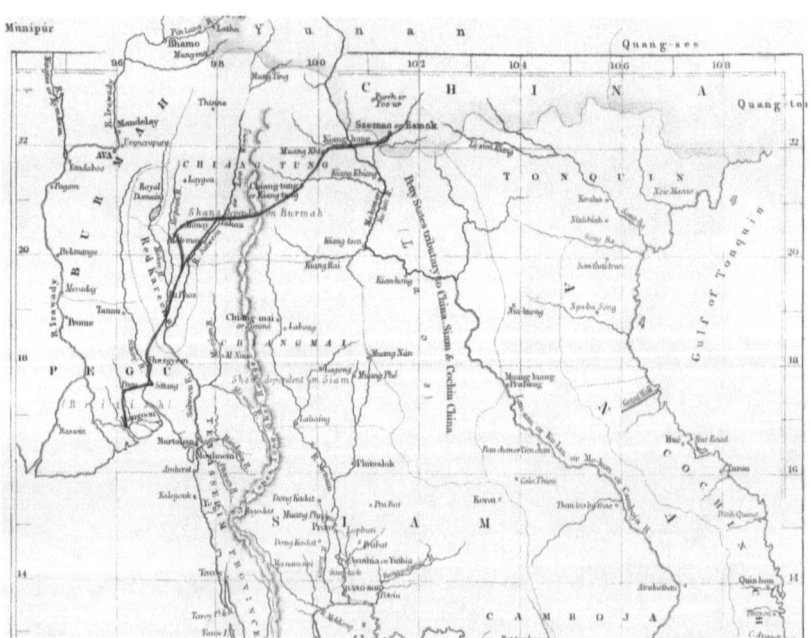

FIGURE 3.2: The Salween with outreaches from the Irrawaddy and Chao Phraya. Source: R. H. F. Sprye, "Communication with the South-West Provinces of China from Rangoon in British Pegu," *Proceedings of the Royal Geographical Society of London, Session 1860–61*, 45.

animals traveling from further inland to Moulmein and other parts of lower Burma and Bangkok. These included an annual flow of 5,000 porters, 3,000 laden oxen, 5,000–6,000 buffaloes, and 200 to 300 elephants. A princess of Chiang Mai, herself a trader, informed Hallett that about 9,000 laden mules and ponies had been coming annually from Yunnan and northern Shan territories including Kiang Tung (Kengtung) and Kiang Hung (Jianghong). This trade route extended from southeastern Yunnan to Chiang Mai before branching off at Uttaradit toward Bangkok and Moulmein, via a combination of land routes and a few navigable points on the Salween. These trading activities were supported by a fleet of 1,000 boats that operated between Labong (Raheng) in Chiang Mai and Bangkok, especially during the rainy season when improved navigability made boat travels more convenient[11] (Figure 3.3).

The next upward major commercial nodal point connected with Moulmein and Chiang Mai was Simao at the southern Yunnan borders. Simao connected Moulmein with Yunnan via multiple roads including Chiang Mai, Kiang Hai (on the Me Khok River), Kiang Tung (the capital of Trans-Salween Shan State), and Kiang Hung.[12] Simao was famous for the Pu'er tea it produced, which was considered "exceedingly rare and precious."[13] The town itself was an attractive inland trading point because of its location as a crossroads of merchant caravans of mules, ponies, and donkeys spanning the routes across Yunnan, Laos, and the Shan States. As Simao was also close to Tonkin, the entire trade route formed a tripartite linkage between the Salween, Mekong, and Red rivers. Further north, the Shan caravanners traded ponies, bullocks, buffalos and cows, silk, boxes, *loongies*, stick lac, stones (such as rubies), and false hair, with a seasonal value of more than 100,000 rupees, and in return, they carried piece goods, copper utensils, and other commodities with a value range of 5,000 to 300,000 rupees (about £30,000 sterling).[14]

At a wider spatial scale, the mule caravans connected these nodal dots that spanned Moulmein and Yunnan. On the return journey from Moulmein, for instance, the mule caravans would carry "every description of cheap cotton goods, bright coloured flannels and odds and ends of trumpery."[15] Although the Salween was largely unnavigable, a combination of fluvial and land networks facilitated connections in varied landscapes, linking people and commodity flows. This rich array of mobilities and interactions, similar to those along the Brahmaputra and Irrawaddy Rivers, inevitably attracted the attention of the British Empire, which sought to establish its presence within this dynamic process.

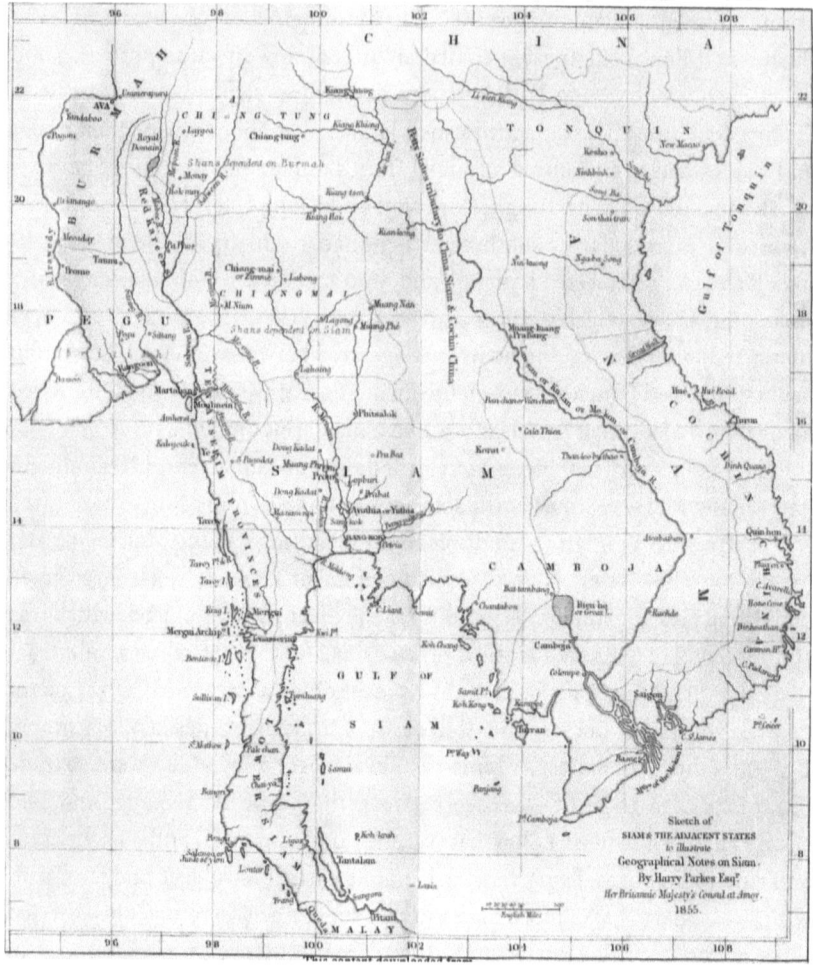

FIGURE 3.3: The Salween network. Source: Harry Parkes, "Geographical Notes on Siam, with a New Map of the Lower Part of the Menam River," *The Journal of the Royal Geographical Society of London*, 26 (1856).

II

The robust river-land trade network of the Salween became a significant site for British imperial encounters, as it sought to establish connections between the Andaman Sea and Yunnan.[16] Immediately after the First Anglo-Burmese War, a British exploratory mission eyed the possibility of improving the Salween navigation to commercially float timbers to Moulmein.[17] Explorations were launched to identify the best trade routes between Moulmein and the

Shan States of Labong, Lagong, and Chiang Mai in a view to further inducing Chinese caravans to Moulmein from Chiang Mai (Figure 3.3).[18] Following the conclusion of the Second Anglo-Burmese War, the British took over Pegu, Prome, and the northern border of Tenasserim, and eyed beyond Chiang Mai to reach out to Simao.[19]

Henry Duckworth, president of the Liverpool Chamber of Commerce, had the BISMRY rivers looming large in his imperial gaze. Beyond Simao, he eyed the seven southern provinces of China: Yunnan, Guanxi, Guizhou, Hunan, Sichuan, Shenzhen, and Jiangxi with a total population of more than 126 million. As China's southwestern frontier provinces had considerable trade with Laos and Annam, Duckworth suggested that the British trade could use the existing Chao Phraya and Mekong routes to connect caravan tracks over the mountain. If this route was not secured, he feared, Russian goods would replace British products in western China using the "highest navigable points" of the Yellow and the Yangzi when the French advance and products from the Mekong was imminent. Duckworth felt that Russia's advance to Asia could be thwarted only by a route spanning Rangoon and Yunnan in which a network of the Irrawaddy, Sittang, Salween, and upper Mekong was crucial. Recognizing the strategic importance of establishing a British emporium on the right bank of the Salween, below the Karen Hills surrounded by its tributaries and accessible to both Chinese and Shan caravans, he wryly remarked that had Russia controlled Pegu, it would have extended its influences from the Bay of Bengal to Chinese frontier via this route "long before now."[20]

Although the Russian threat was unlikely to materialize in this region, an excuse of possible Russian expansion was needed to argue in favor of an early British connection across Arakan, Moulmein, Pegu, and southern Yunnan along the Salween basin, especially the Shan States.[21] The British exploration that followed in this direction, however, did not bring the desired results. The situation was fraught, not only due to the uncertainty stemming from some Shan chiefs rebelling against the Burmese king but also because the Burmese government itself adopted a dual stance. Burmese officials welcomed the British exploration team with traditional gifts and permission to proceed, but secretly ordered the villagers not to sell any provisions to them. The Burmese were worried about losing their remaining political and commercial clout should the British establish their authority along the Salween trade routes that connected China and Moulmein. This Burmese bout of resistance, although short lived, looks similar to what Cooper experienced from the Mishmis in

Sadiya in the upper Brahmaputra and Sladen from the Kachins in the upper Irrawaddy. The exploration along the Salween in 1864, led by Lieutenants Watson and Scone, ended with the high hope that the establishment of a bazaar on the British border would bring in Shan migrants and eventually "open out a means of commercial adventure to the more distant Shan States to the north, and the Chinese frontier Province of Yunnan."[22]

A more vigorous advocacy for trade connections between southwestern China and the Gulf of Martaban along the Salween network came in the 1870s from John Coryton, a record keeper at Moulmen. Coryton opposed an earlier plan for a railway route from Calcutta to Canton along the Brahmaputra in anticipation of challenging engineering work and high expenditure. He was equally opposed to the option of connecting southwestern China and Rangoon via Bhamo on the Irrawaddy on political grounds since upper Burma was still under the Burmese king. His focus was on the Salween, which ran "precisely in the direction of the traffic we are desirous of attracting."[23] He asserted that "every ounce of traffic" that went from western China to Rangoon "must cross our river—the Salween."[24] Coryton saw that Moulmein had the support of "Nature—an ally in the long run more powerful even than the patronage of a Local Government." He argued that compared to the Rangoon-Bhamo route, Moulmein-Chiang Mai route was only one-fourth of the distance at about three hundred miles. He contended that "nature often rough hews a purpose and leaves the ends for man to shape" as he envisioned a vast system of internal water communication across the Salween, Yangzi, and the West rivers spanning Moulmein, Shanghai, and Canton. In this water network he saw great potential for investment in trade and commerce, for he thought that capital was "timid but not stupid."[25]

Coryton's vision of expanding imperial trade via a land-water network was supported by his colleagues. But the colonial administration could hardly bring any remarkable change to the existing pattern of mobility. Although the administration had launched several exploratory expeditions, their efforts to establish trade routes along and across the Salween valley met with substantial resistance from various local groups. If the imperial wheel slowed down on the Brahmaputra above Sadiya and on the Irrawaddy above Bhamo, on the Salween this happened much lower down the river. Part of the resistance was from the last vestige of the Burmese Empire, as noted earlier. After losing the middle Irrawaddy and with the constant threat to its existence from the advancing British on the remaining upper Irrawaddy, the Burmese attempted unconventional defense in the Salween corridors. One way of discouraging the British explorations was to frighten them about the danger posed by the non-Burmese population on the northern route, as they did to Sladen when he took the Irrawaddy route to reach Yunnan. When a team wanted to go up the Shweli River, they were warned about the danger from "half-wild" Kachins, and that warning indeed prevented the British team from reaching Bhamo via the Shweli.[26] However, these Burmese techniques for containing the British were only temporarily successful, as the Third Anglo-Burmese War dashed any hopes of reviving the faltering Burmese kingdom.[27]

The overthrow of the Burmese king, however, had little impact on the extent of British power, as the Chinese presence grew more immediate with the final collapse of the remaining territories of the Burmese kingdom. As disputes about the Burma-China frontier demarcations intensified, the upper Salween became a focal point in the 1890s. Chinese negotiators began to assert their rights over the upper Irrawaddy above Bhamo, on the Mali Hka and Nmai Hka, and crafted an air of naturalness about China's rights over these territories, suggesting that the Burmese claim in the headwaters of the Irrawaddy was being heard for the "first time." Despite such assertions, Chinese negotiator Macartney suggested that the demand for a share of the territory east of the Irrawaddy could be withdrawn if sufficient compensation on the Salween was provided to China.[28] This stance was a preemptive assertion on the Irrawaddy to secure China's claim on the Salween. Eventually, the Chinese diplomatic efforts prevailed. The Pianma expedition of the British in 1910 across the Yunnan border was launched to bring under their control the watershed of the Nmai, Shweli, and parts of the Salween as far as latitude 26°15′N as well as the valley of the Ngawchang Hka. It seemed, however, that the British were "successful" against the Chinese only in securing territories that were not doubtlessly part of Burma. As British officials conceded, the Chinese claim to the upper Irrawaddy watershed was "not a very strong one, yet it is stronger than ours, for we have none."[29]

After World War I, Anglo-Chinese conflicts resurfaced, and this time too the British forces were unable to fully utilize their power across the northeastern frontier with China. Maintaining an Anglo-Burmese army near the Chinese border required transporting all necessary supplies, but the available infrastructure was inadequate. In fact, fortified posts in the Salween Valley relied on meat and other provisions from garrison cities in western Yunnan, effectively making enemy territory a key supplier for the British military.[30] The Anglo-Chinese border tension eased somewhat as the Japanese threatened both in the 1930s, although the Chinese continued to assert their claims over the riverine tracts west of the Salween.

The high politics around the Salween, involving the British, Burmese, Chinese, Siamese, and latterly the Japanese, resulted in no single power's exclusive sway. This was not only due to shared challenges of unfamiliar terrains of the Salween basin but also because the local inhabitants, operating within their own ecological settings and along vital trade routes, were a significant force. Although the British occasionally launched "punitive expeditions," most policies were clearly aimed at securing the Salween trade network through the appeasement of local inhabitants. This approach mirrored the traditional Chinese method of ensuring access to the Gulf of Siam via Chao Phraya river or the Gulf of Martaban via Moulmein. Both the British and the Chinese had to contend with the ethnic communities inhabiting the Salween network, including the Karens and the Shans. Deeply rooted in their local environment, the ethnic groups vigorously protected and promoted their interests in the flows of commodities, humans, and pack animals that the Salween elegantly facilitated.

III

The unnavigability of the Salween, which began close to its mouth, forced the British Empire to engage with the local ethnic groups much further down the river compared to other BISMRY rivers under their sway. The terrain was mostly rugged, forested, and fluid, yet the imperial need to maintain commercial connections up the Yunnan was strong. However, British knowledge of the region during the late nineteenth-century imperial advances was scanty, to say the least. Coryton refers to a competent British Indian official whose "some small error" excluded a thousand square miles of territories from British sovereignty.[31] All these necessitated the application of the skills, knowledge, and network of the local inhabitants along the river basins. The outcome was a mix of compromises and conflicts, which eventually shaped a shared space for mobility that bonded Yunnan highlands with the Gulf of Martaban and adjacent areas.

Living close to Moulmein, the Karens were at the receiving end of commodity flows in the Salween network, as well as from the river networks in Siam to the east (especially the Thaungyin Chaung or Moei River) and the Yunzalin River, a Salween tributary, to the west.[32] The Karens believed that their long journey over two millennia ago from the Gobi Desert ended in the lower Salween, suggesting that the entire stretch of the Salween was part of their collective memory. The Yunzalin was navigable by country boats as far as Papun, the headquarters of the Salween district. The Salween was generally navigable by country boats within the district (except at the Hatgyi, the Great Rapid).[33] In other words, the Karens occupied all the streams and brooks and controlled the commodities that flowed through and were consumed in their neighborhoods. Some worked as coolies, capable of carrying up to seventy pounds over the steepest hill "without any appearance of being overfatigued."[34] Describing the Karens living in the hills, British military official McMahon notes that they were "quite as unmanageable" as the Kachins, but were aware that their profits depended "on the amount of traffic that passes through their country, and they will be only too glad to offer all facilities for trade if it be made worth their while."[35]

If the British and the Karens had apparently developed a mutually expedient relationship, it was in the teak extraction region, north of the Karen territories on the Salween, where confrontations intensified. Behind each log that landed in Moulmein for shipment, there was a complex story of imperial encounters with local inhabitants. Some allowed the cutting and drifting of logs in exchange for profit, while others resisted in various ways. Regular theft and robbery involving people in the teak business was considered a law enforcement issue.[36] Behind most of these occurrences, however, there was a network of people who considered petty crimes as a form of resistance against external encroachment into their ecosystem. Many had connections with the Karenni leaders, whom officials branded as *dacoits*. In the 1870s Coryton noted "all quiet in Karenne" as the *dacoits* on the east bank of the Salween dominated the region while watching over the *thitgoungs* (foresters).[37] One such *dacoit* leader was identified as Moung Deepah, a "celebrated Dacoit," as British explorer Colquhoun described him.[38]

The elephant played a crucial role in teak transport (Figure 3.4). On one occasion, Deepah aimed to recover the elephants confiscated by the colonial administration. On another, his group confronted the cutting and trading of teak in Karen and Karenni territories. Various strategic points on both banks of the Salween and its tributaries became Deepah's field of operations, where his spies monitored British forces, making it "next to impossible to make any movement with secrecy" and to protect about 150 miles along both banks of the river. The legitimacy they enjoyed was clear from the fact that Deepah's

FIGURE 3.4: "Hauling logs from the Salween River with the patient, powerful elephant," Moulmein, Burma, 1907. Source: Library of Congress, https://www.loc.gov/item/2019632198/.

boat was decorated with gold leaf, symbolizing his authority to collect timber revenue on behalf of the Eastern Karenni chief and to confront the British forces if necessary.[39]

The resistance of the Karennis hardly stopped the teak trade during the last decades of the British rule in Burma, but it did slow it down. From the British perspective, the Karenni resistance was more than a localized challenge, as they guarded trade routes to China. Overcoming this challenge depended on the cooperation of other "friendly" ethnic groups, bringing them closer to the Shans who were demographically larger and more widespread.

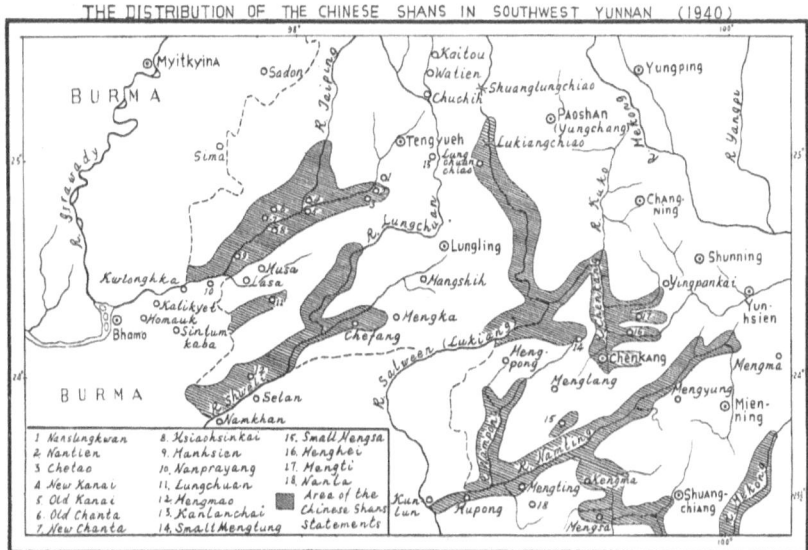

FIGURE 3.5: Riverine Shan settlements in southwestern China. Source: Yin-Tang Chang, "Anthropological Features of the Shans and Their Geographical Environment in South-West Yunnan," *Man* 44, no. 55 (May–June 1944): 62.

The Shans occupied major river networks, spanning many tributaries of the Irrawaddy, Salween, and Mekong (Figure 3.5). These included Taping (a tributary of the Irrawaddy) all the way to Tengyueh, the Shweli valley, the Salween's major tributaries of Wampong and Namting, and part of Mekong (below Mien-ning) and to a lesser extent the upper Red River Valley. They usually inhabited the lower plains of the river valleys at altitudes of four hundred to one thousand meters above sea level.[40]

As Shakespear noted, they were found "everywhere" from Dali to Assam to Bangkok, and at the time of the British arrival, they were settled in disparate places such as one hundred miles north of the Mgoung, Hukong, and Tanai valleys where the Chindwin River originates.[41] In these strategic locations, the Shans covered waterways, valleys, and other trade routes connecting most of mainland southeast Asia, Yunnan, and the northern Brahmaputra basin. The British, who eyed uninterrupted trade with China along the Salween network, could not have found a better partner than the Shans.

The imperial depiction of the Shans fit their commercial potential and partnership along the Salween routes to Yunnan. MacMahon underscored the Shan commercial acumen, reflected in their mobility between Burma and China, and highlighted their frugality, industriousness, perseverance, aptitude for business, and "trading proclivities of the Chinese." He noted that they carried their goods "across seemingly impassable hills, and extending trade in all directions by every means in their power."[42] Even the Shan bullocks paid "little regard to the character of the road as long as a march or two is saved."[43]

The Shans, on their part, met the British imperial expectations as trade partners. While the Karens operated within their comfort zone, the Shans pushed their boundaries in and beyond their usual habitats. Their "almost instinctive recognition of time, direction, distance," as Coryton noted, made them "admirable traders," which was aided by their "love of wandering life."[44] Additionally, the Shan traders and travelers were considered skilled mapmakers, creating relief maps from their firsthand knowledge while gossiping over tea. R. G. Woodthorpe, an explorer of the Chindwin River, found the scale of Shan maps to be a day's march to a notched stick, and noted that despite the roughness and unscientific method, the results were "useful." Traversing the territories across southern Tibet, western China, and their own Shan States, they could prepare maps from the vantage point of any trade center, such as Darjeeling, Dibrugarh, Batang, Dali, Bhamo, Luang Prabang, and Chiang Mai, as Woodthorpe noted.[45] These soft skills were matched by the Shan physical prowess, being splendid walkers, mountaineers, and foresters. Younghusband noted how a group of Shans kept pace with their mule caravans carrying heavy loads along the bumpy roads from Chiang Mai to Kiang Tung, averaging over twenty miles a day.[46]

These mobility skills were matched by another aptitude for making bamboo bridges, some of which were strong enough to sustain motor cars and lorries.[47] The Shans also impressed the colonial observers as producers and consumers of crafts and as traders themselves. "Tasteful Shan ladies, with their love of jewellery and elegance of design, may yet give a fashion to Belgravia; and the Shans, being famous workers in straw, may yet produce hats for the park rivalling the Tuscan in quality, and in shape that of the far-famed beautiful 'Duchess' or excellence in the manufacturing of cotton cloths."[48] As consumers, the Shans opted for both cheaper calico German and Manchester goods as well as expensive German and Belgian finery and hardware, so much so that European traders took special care to conform to these consumers' needs.[49] Considering the commercial acumen of the Shans, Hallett went so far as to suggest that they wanted "free trade."[50]

Beyond their natural proclivity for trade and commerce, the Shans were advantaged as translocal traders by their relative immunity from malaria, which the Salween Valley was infamous for. Since the time of Marco Polo's travels, Europeans have viewed the Salween Valley as a "valley of death," a perception echoed by the Chinese. For example, Mengting had an annual rainfall of 1,644 mm accompanied by intense humidity and heat, with temperatures above 20°C for eight months, making it prone to malaria compared to the relatively healthier Kunming, which saw about 1,245 mm of rainfall. Yet, at Mengku, only 30 miles further north, the Chinese were "flourishing," as a report observed.[51] The notoriety of some places along the Salween, such as Tangyan, was proverbial, as reflected in a Chinese saying: "Before going down to Tangyan, you should marry off your wife to another man."[52] The Shans appeared to have been able to evade the ferocity of malaria and miasma through millennial acclimatization.[53] Their relative immunity to the tropical disease and their extraordinary scale of transregional mobility left the Shans not only comfortably placed along the Salween corridors but also in control of commercial nodal points of a few BISMRY rivers.

As the British faced the formidable Karenni beyond the territories of the friendly Karens, they also encountered the allegedly intimidating Wa people beyond the territories of the agreeable Shans. The relationship between the British and the Was, who mostly lived on the hilly basins of the river, resembled that of the British and the Abors at the headwaters of the Brahmaputra. The British could hardly tolerate them, but could neither afford to disengage as many roads to Yunnan fell within their territories. When the British reinstated the Wa Chief Tonhsang as the *sawbwa* (chief) of East and West Manglun in 1891, he was required to pay monetary tribute for West Manglun and an annual *nazr* (gift) of gold and silver flowers, among other items, for territories east of the Salween. However, Tonhsang did not see the Salween as a dividing line or boundary. He claimed his territorial rights on both sides of the river, and asked for a symmetrical tax assessment. J. G. Scott, superintendent of the northern Shan States, proposed a lump sum of 500 rupees for 1892–1893 for the entire Wa region, suggesting that the government could better lower the tax for Tonhsang's trans-Salween possessions and demand 300 rupees only for West Manglun for that year. He recommended demanding "nothing beyond the pony, dried squirrels, &c already presented by way of tribute" and proposed remitting the tribute for the previous year's tax altogether.[54]

British officials had little choice but to be lenient, as the Wa rulers, under Chinese patronage, occupied a strategically advantageous position between the British and the Chinese. The Chinese administration conferred such status as "aboriginal major" to local chiefs. The British administration attempted to emulate the same practice, as they did with the Abors and Mishmis on the Brahmaputra and Kachins on the Irrawaddy.[55]

In practice, the Was could not be bribed into total loyalty as much as the British or the Chinese desired. The British realized that the Was couldn't be "divided" between the British and the Chinese and planned to exclude them as a whole from Chinese control by drawing a boundary around the Kongmin or Loi Maw range, which formed the eastern watershed of the Nam Hka and divided the basins of the Salween and the Mekong. These boundary lines were drawn with the gold mines at the source of the Nam Hka in mind. However, most Wa chiefs were determined to keep out the Chinese, while being equally opposed to the British rule, and did not allow the extraction of gold. In the end, the local government suggested that until there was a settled frontier with China, the Was had better be left alone.[56]

In the Salween network, the Karens and the Shans were eager collaborators in trade with the British. The Karenni and the Was held suspicion, but they were equally eager to participate in trade activities. On their part, the British officials needed not only to acknowledge their presence along the Salween basin but also to appease them. For example, Scone, an early British explorer, made concerted efforts to gain the trust of various ethnic groups between Pegu and Yunnan, offering medical aid and engaging in open personal interactions with local chiefs.[57] The extent to which these outreaches promoted a common space of engagements at the Salween networks remains a subject of further exploration, but certain elements of entanglement were favorable to such a possibility.

IV

Each ethnic group inhabiting the territories along the Salween network had specific interests to serve, brewing tensions and sometimes full-scale conflicts. However, the intense mobility and flows made one thing almost inevitable: interactions and assemblages. In the Amherst district where the Salween met the Gulf of Martaban, a "consortium" of ethnicities included the Talaing, Karen, Burman, Shan, and Taaugnthu as well as people of different backgrounds from India, China, and Europe. While a majority of these groups concentrated at

Moulmein, seasonal travelers like Kha people from as far inland as Luang Prabang on the east bank of the Mekong came to assist the Moulmein teak merchants, adding a riparian vibe to a maritime vivacity.[58] In the early 1840s, Moulmein hosted a plethora of international merchant establishments, including eleven "Mughals" (possibly meaning north Indian Muslims), seven Burmese, six Parsees, two Armenians, fourteen Chinese, three Surati, one Hindu, and three Jews—as reported in a contemporary magazine.[59]

The cosmopolitan synergy at Moulmain, like the commercial flows, extended to Bangkok. This led to the flourishing of several nodes, including Mae Sot on the Burma-Thai border, which in turn was connected to Chiang Mai further north and Bangkok to the south. Traders from Yunnan, the Shan States, Laos, and northern Siam transacted with Indians, some of whom had migrated to this region following the mutiny of 1857. The migration persisted as Indians relocated from Burma following the Japanese invasion, with approximately 30,000 individuals, including around 2,500 Bengalis, residing in Chiang Mai by the 1970s. Bengali Muslims, in particular, played a crucial role as intermediaries between Indian migrants and mainland Southeast Asian communities. Eventually some of them moved upward to Chiang Mai and others to Bangkok.[60]

Soonthornpasuch traced one of the earliest known Bengali settlers, Usaman Miyashi. His parents were originally from Calcutta and migrated to Eastern Bengal and then to Rangoon. Miyashi was born and brought up in Rangoon and took up business, finding himself traveling extensively between Rangoon, Moulmein, northern Thai border towns of Mae Sod in Tak province, Mae Sariang in Maehongson province, and Hod in Chiang Mai province and Chiang Mai city. Miyashi married a Burmese wife, settled in Chiang Mai city with his family, and became the headman of the Muslim community of the region. He also became the first imam of the mosque built on the bank of the upper Chao Phraya (Ping) River. The land for the mosque, granted by the prince of Chiang Mai, was located on the river that connected Chiang Mai and Bangkok with Moulmein on the Salween. In this area, the Bengali community led by people like Miyashi shared the migrant space with communities such as the Malays.[61]

Further upstream, the Salween's confluences with smaller rivers occurred at shorter intervals, leading to concentrated populations and flourishing markets, although on a smaller scale compared to Moulmein or Bangkok. In the Karen territories of Yembine, situated at its junction with the Salween, settlements of multiple ethnic groups developed in connection with natural resources of the Salween basin and opportunities for commercial transactions. For example, in Yembine and adjacent Chao Phraya network, there were 33 Lawa villages, 46 Karen villages, and 11 Shan villages, representing about 30,000 people living in a mixed ethnic environment. The Karens maintained barter trade with both the Shans and the Yunnanese caravan traders, who brought silk thread, straw hats, and copper pans and purchased salt, betel nut, and other goods along the way.[62]

Other parts of the Salween Valley were dotted with inclusive villages of Shans, Kachin, Wa, Musho, and Chinese. Although different ethnic groups showed a propensity to certain elevations—such as the Kachins and the Was living in higher zones than the Shans—most of these groups were connected by trade relations.[63] Especially at the Salween's confluences with its tributaries, such as the Meh Kha or Meh Li rivers, everyday commercial transactions thrived alongside large business conglomerates like Chinese Hong and Buddhist monasteries. Large villages were also located at these confluences, which were locally known as the *pak*.[64]

Some ethnic groups opted for rooted mobility, remaining relatively place-bound but connected to more mobile groups like the Shans. In contrast, other groups continued to move between and across river valleys on a larger spatial scale. Ta Fu Yeh, the spiritual leader of Lahu community in Mong Hka, told J. G. Scott that the Lahus had originated from the sources of the Irrawaddy and Salween. They were once obliged to accompany a Burmese king to China to obtain the tooth (*swedaw*) of the Buddha and subsequently settled on the Salween valley of Mong Myen. Later they were driven by the Chinese, leading them to migrate to the south along three routes: one group moved along the Mekong, another crossed the Salween and settled in His Paw and Hsen Wi, and the third settled in Nan Cha (Ho Sak as known among the Shans). At the time of Scott's reporting in 1893, the cluster that had settled along the Mekong was still moving southward into Siam, even beyond Chiang Mai.[65]

Those who inhabited the eastern Salween bank, a thirty-day journey northwest of Kiang Tung, had their capital in Koo-lie Muang Khan, situated at the head of the Meh Kha, a Salween tributary. They cultivated glutinous rice, tobacco, cotton, and chilies, bartering the surpluses with the Shans and others.[66] The territories further north, where all of the BISMRY rivers offered multiple routes for commercial mobility, paved the way for similar social interactions. The Lisus, for example, lived mainly on the west bank of the Salween in a location almost equidistant from Lijiang, Sadiya, and Bhamo. It was not unusual to find a Yunnan merchant married to a Lisu with daughters, seeking suitors who would bring trade and profit to the family.[67]

The Tibetan ethno-commercial connection to the Salween was strong in the upper flows in Yunnan. Desgodins describes how Tibetans from Tsarong exchanged salt for grain with the Nung, a non-Tibetan ethnic group in the Salween Valley. The Nung transported five hundred to six hundred loads of salt annually from Tsarong, trading it at a ratio of one load of salt for every five loads of cereals.[68] The Tibetan connection to the Salween trail thinned further south, but the spiritual connections and flows remained alive. Around latitude 28°20′N, a sect of heterodox Lamas traditionally consigned their ritual paraphernalia to the waters of the Salween as part of their religious practices. When asked about the destination of these offerings, they would respond, "What will become of it all? The great river, whose waves roll to Martaban, is no more than two or three hundred paces away."[69] This response suggests that, in the minds of the Lamas, Tibet and the Indian Ocean were spiritually interconnected, with the Salween serving as a conduit between them.

In the rugged terrain of the Salween Valley, where pathways ran several hundred feet above the river, villages were still established nearby. Their proximity to the river was essential, as they relied on the small tributary streams that fed into the Salween for irrigation and daily necessities.[70] Human settlements, in the form of villages and towns, were "numerous." However, in terms of size, a bazaar along a confluence could be larger than a town. This was because the market not only served local needs but was connected to translocal nodal points along the Salween network. For example, on a specified day in the bazaar on the outskirts of the small town of Tulay, located on a large stream that runs into the Mobyai River, five thousand people from different regions used to come on boats.[71]

The steady ethnic mobilities and entanglements in the Salween basin were also reflected in the *zayat*, an institutionalized facility for travelers and traders, featuring simple yet varied architectural designs (see, for example, Figure 3.6). The *zayat* was a simple roadside rest house, customarily made available for travelers by local inhabitants.[72] Ellen Thorp, an English author who grew up in Taunggyi in the western Salween basin, described two types of *zayat*s. One was a traditional local hut, raised on mud piles, with mat walls, a thatched roof, and a split-bamboo floor, which looked "clean and comfortable." The other kind was a makeshift one, usually made of a yak-hair tent. The *zayat*s

FIGURE 3.6: "Zayat and Temple in Burma." Source: John Baillie, *Rivers in the Desert: or, Mission-Scenes in Burmah* (Seeley, Jackson and Halliday, 1859).

were erected at regular intervals along the travel paths and often around a confluence or a tributary of the Salween. Their presence often shaped a built environment "like a beautiful stage-set; nearby a glinting stream flowed, behind the thatched roof purple shadows fell athwart group of banyan trees standing between the zayat and the road."[73]

As a place for rest and gathering, with a touch of a local tradition of hospitality for people on the move, the *zayat* was a signifier of mobility that kept the Salween trail alive. It often found itself in the middle of larger social and cultural entanglements near a pagoda or marketplace that connected the Salween trails. For example, Tailan *zayat*, situated on the bank of a stream of the same name, was a site where the Shan and Karen caravans generally rested with as many as one hundred bullocks.[74]

In the Salween corridors and networks, mobility and multiracial entanglements in a riparian environment were normal affairs, despite the tough physical terrains. At the Meung-Keung festival, where people combined "piety with business," Thorp encountered a "sea of humanity" that included the Shans, Palaungs, Chinese, Taungthus, and Burmans, among others. The environment was vibrant: "Naked children cried and laughed and sucked sticks of

FIGURE 3.7: Monkeys and a crocodile at play as humans pass by on the Mekong [In Chantaboun]. Source: M. Henri Mouhot, *Travels in the Central Parts of Indo-China*, vol. I (John Murray, 1864).

sugar cane. Crippled folk and blind knelt before the flickering candles. Pariah dogs snapped and fought for scraps."[75] Thorp noted the Chinese, Indian, and Burmese shopkeepers offering "a wonderful medley of merchandise. Silk from Japan, cotton goods from Manchester, toys from Germany, earthenware and enamel goods, mirrors covered with gaudy flower paintings, glass bangles and necklaces, needles, cotton, scissors and biscuits, sugar, bags of flour, rice and occasionally a few tinned foods." She described the liveliness of Ishar Singh's tailoring shop, the colorful pictures of Hindu gods and goddesses in Takaram's store, and similar other scenes. One evening, within this lively assemblage of people, commodities, and other-than-human entities, Thorp heard both church and pagoda prayer songs, which gave her a "sense of the universal."[76] This feeling of the universal was reminiscent of Hallett's experience a century earlier while on a boat in a Shan territory at the confluence of the Chao Phraya and Meh Li on the way to Chiang Mai. Immersed in nature's "music, life and light," he felt a "universal" enjoyment that "wafts care away, renews our youth for the time, and we enjoy the pleasures of paradise"[77] (see a similar idyllic depiction, on the Mekong, in Figure 3.7).

~~~

Some imperial officials in Burma observed that the Salween, with its narrow, rocky bed and roaring rapids, was unsuitable for full-scale navigation. However, they acknowledged its significance as a waterway, considering it "of no less value than its eastern sister, the Mekong."[78] The colonial state's designation of the Salween as an unnavigable yet important waterway shows that it was caught between the challenge of the unnavigability of the river and its existence as a useful track. Under these circumstances, while maintaining a perception of the river's uncertain utility, the administration made commercial use of it by floating timber downstream to Moulmein and facilitating bank-to-bank ferries for traffic. In the process, it had to negotiate with riparian ethnic groups regarding the forested and rugged landscapes alongside the river. Some local ethnic groups retained the exclusive right to work as ferrymen.

Although the immanent power of the unruly river on its bed as well as on its banks meant that the British Empire had to be content with uncertain extraction, the local inhabitants capitalized on the same navigational difficulty to contest the imperial incursions. At the Salween, therefore, the empire encountered the combined agency of human and nature. Rather than hindering its development, the frictions, contestations, and negotiations between imperial forces and local populations contributed to the Salween's evolution as an ecological commons. This space became accessible to multiple stakeholders, extending from the Gulf of Martaban to Yunnan and encompassing neighboring rivers, including the majestic Mekong, to which we now turn our attention.

FOUR

MEKONG

Rendezvous with the Mother of Waters

THE MEKONG EVOKES MYRIAD meanings, shaped by its landscapes and histories, and the visions of those who encounter it.[1] During a survey of Yunnan in 1902, George Litton, a British consul in Burma, imagined the Mekong as a "huge conduit pipe" through which the waters of Tibet flowed to the South China Sea.[2] This idea resonated with the French colonial perception of the river as "singular and remarkable," suggested a few decades earlier by Francis Garnier, a leading French explorer of the Mekong.[3] The European vision of the Mekong as a linear and exclusive body of water perhaps stemmed from their relative lack of access to the river because of its unnavigability, which foreclosed the possibilities of lived intimacy between the empire and the river. Indeed, it was declared "comme voie commerciale presque inutile" (an almost useless trade route).[4] The imperial experience with the Mekong, therefore, became, as Milton Osborne lamented, synonymous with the transition from hope to disappointment and from challenge and illusion to defeat.[5]

For those living in the Mekong basin, the river was more than a conduit pipe; for them, it was simultaneously a crossroads, a terminal, and a track. The Mekong's tributaries and their confluences with it, along with the zones of micronavigability it provided (as in the case of other BISMRY rivers), created a network of mobility that mattered more than through navigability of the river between Tibet and Saigon. Because the Mekong was mostly unnavigable, its banks became a more pronounced site of mobility and served as springboards of connections with other river valleys, especially the Salween in its upper riparian regions and the Chao Phraya system in the southwest—defying the perception of the Mekong as a lifeless tunneled flow of water. This understanding of the Mekong as a multifaceted lifeline, rather than just a body of water, sets the stage for exploring its broader historical significance.

I

Where the Mekong parted from its running mates in Yunnan, the Salween and the Yangzi, it was merely a stream (*ta-kiang*) rather than a fully formed river. Between Dali and Simao (also spelled Szemao or Sumao), the river was typically less than 200 yards wide, with two boats often forming a makeshift bridge for crossing. In this stretch of the river, there were more than 16 crossing points linked different nodal areas, with Simao, 40 miles east of the Mekong, being the most prominent. Simao was connected to Chiang Hung on the Mekong by a six-day caravan journey. It was halfway between the Kunlon Ferry over the Salween and Manhao, the starting point of boat navigation on the Red River; and was also close to at least three small rivers that connected the Black River, a major tributary of the Red. In other words, as the crow flies, Simao was fairly accessible to multiple nodal points like the Kunlon Ferry on the Salween, Mengzi near the Red, and Dali on the Mekong[6] (Figure 4.1).

Simao's location at the crossroads of several BISMRY basins turned it into a busy commercial hub, especially during the nineteenth century. Imported goods in Simao included Manchester woolen and cotton piece goods, blankets from the Netherlands and Turkey Red from Switzerland, lamps, soap, mirrors, medicines (Davis's Pain Killer, Elliman's Embrocation, and a toothache remedy), bird's eye tobacco, and wax candles. Most imports, which arrived in Simao via the Irrawaddy, were destined for the Chinese interior, transiting through Hunan, Hubei, and Jiangxi. Exports to Burma via the same routes included silk, opium, and white wax, among other products.[7]

Simao's importance as a commercial hub was bolstered by its proximity to the town of Pu'er, famed for its namesake tea. Pu'er enjoyed an estimated yearly output of 15,000 mule-loads, or about 900 tons, with a value of 670,000 rupees (£45,000 sterling).[8] The urban development resulting from these transregional flows did not escape the attention of European explorers like H. R. Davies, who compared Simao, on a smaller scale, with New York City in terms of commercial activities, and Kunming with Washington, DC, with its "roomy houses, gardens, and wide streets."[9]

Another major nodal point on the Mekong was Chiang Hung, located in the Shan heartland of Xishuangbanna. It was a crossroads for commodities

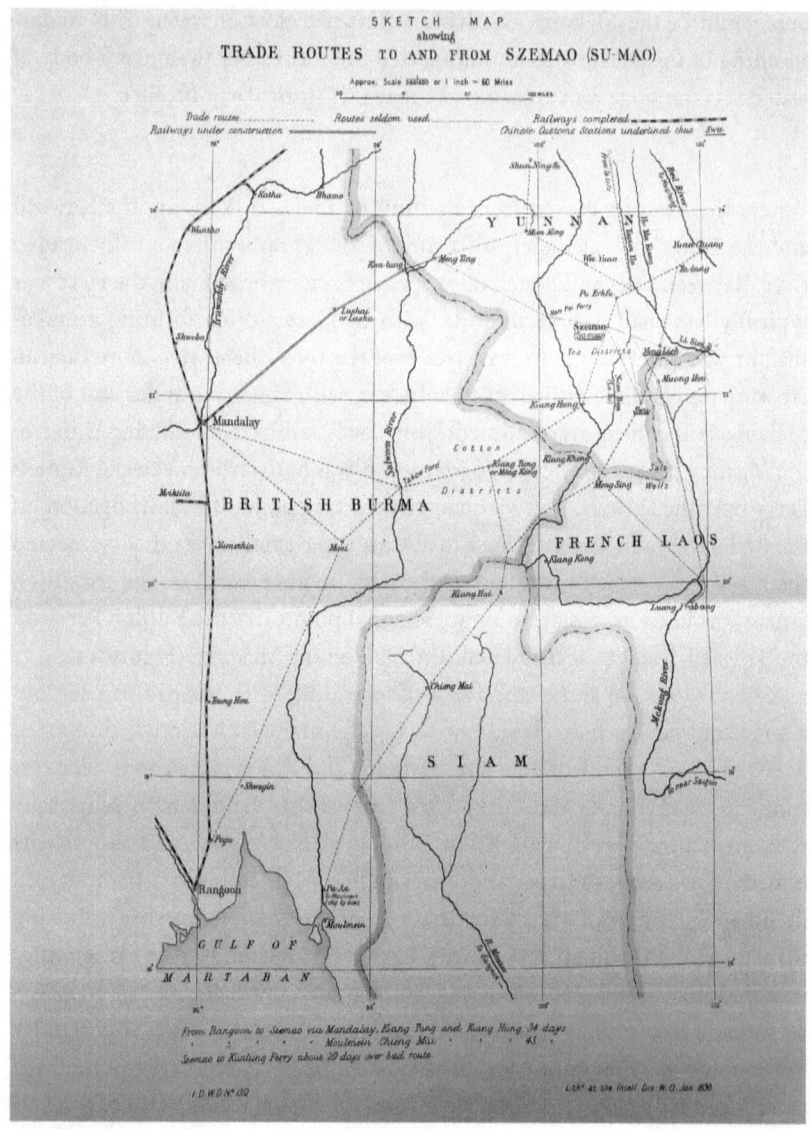

FIGURE 4.1: The Simao trade network across the BISMRY valleys. Source: London, The British Library, L/PS/18/B113: Trade of Szemao (Yunnan) with Burma. Its commercial importance. Mandalay-Kunlon Railway and its terminus on the Salween.

and people from Dali and Kunming as well as from the Red and Black River Valleys via Pu'er and Simao. Situated on the right bank of the Mekong, Chiang Hung was also closer to the Salween network. Here, the Mekong was more than five hundred feet wide and fifteen feet deep, navigated by boats of various sizes even at the driest time of the year. Caravans from China crossed it in ferry boats, with this caravan track being dubbed a "Great Route."[10]

From Chiang Hung, one route led southwest along the Salween via Kengtung, while another sprang due south along the Mekong to Chiang Saen in Luang Prabang. Andrew Walker noted that despite rapids, sandbanks, and rocky shoals, the river was "navigable in all seasons" between the Lao provincial capital of Houayxay and Luang Prabang for boats carrying up to 120 tons. At Luang Prabang, however, the Mekong traffic tended to shift more toward Chiang Mai than toward its lower stream in Saigon. This preference for westward land routes off the Mekong was due to lack of thorough navigability below Luang Prabang and the relatively quicker access to the Indian Ocean via the Chiangmai–Bangkok route, facilitated by the Chao Phraya network. Consequently, Chiang Mai became a pivotal destination for the swinging Mekong traffic, as discussed in the previous chapter.

Between the banks of the Mekong and the Salween, from east to west, four major tributaries of the Chao Phraya flowed near Chiang Mai: the Nan, Yom, Wang, and Ping Rivers (Figure 4.2). These rivers played a key role in connecting both Moulmein and Bangkok with Yunnan.[11] An example of these rivers serving as commercial arteries is the Lagong on the Wang River, which facilitated significant traffic from Yunnan via the Mekong. Nine or ten large Chinese boats commuted every month between Bangkok and Lagong each, bringing goods worth 90,000 rupees to Bangkok. These goods arrived in Lagong from Yunnan in Chinese caravans of 30 to 80 mules. The city was also visited by about 10,000 laden cattle and 20,000 porters from neighboring areas on their way to Bangkok and Moulmein.[12] In other words, Chao Phraya's northern headwaters had natural outreach to both the Salween and the Mekong, making it well positioned to mediate the trade flows between Yunnan and the Indian Ocean.

The commercial flows that continued along the Mekong below Luang Prabang added to the nodality of Phnom Penh and Saigon. Phnom Penh was situated on the mainstream Mekong but was also surrounded by an intricate network of waters, including the Tonle Sap Lake, which draws a cluster of tributaries. Together, these features lent the region the shape of a fluvial crossroads. Near Phnom Penh, the Bassac River, the Mekong's largest branch, formed a confluence and marked the start of the Mekong Delta, giving rise to a series of distributaries collectively known as the 'Nine Dragons.'[13] Further down, on a branch of the Mekong, Saigon has stood out

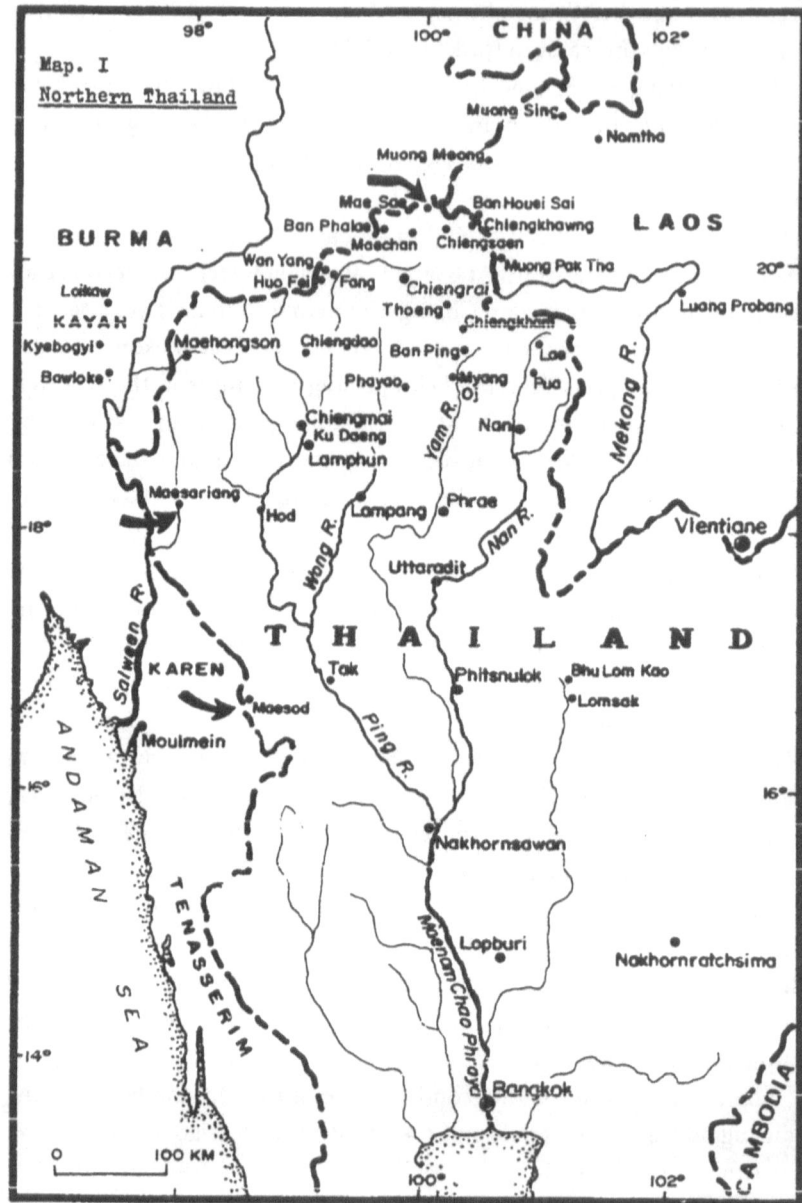

FIGURE 4.2: Mekong outreaches to the Chao Phraya and the Salween. Source: Suthep Soonthornpasuch, "Islamic Identity in Chiengmai City: A Historical and Structural Comparison of Two Communities" (PhD diss., University of California, 1977), 10a.

as a mega signpost to and from the South China Sea since the mid-nineteenth century.

Saigon, in turn, generated a circular flow of trade along the coast of Annam and up to Tonkin. Paul Macey traveled from Saigon along the coast of Annam to the Red River and returned to Saigon by the Black, Nam-Hou and Mekong. The trip took over seventy-five days, of which seventy-one were spent on waterways and the rest on ox-cart or horseback.[14] This experience of circular mobility through a river-ocean-mountain maze between the Red and Mekong was also described in the travel account of Charles Cochrane-Baillie (known as Lord Lamington), a former governor of Bombay and Queensland.[15] The circularity, partially triggered by Saigon, is also reflected in the trade statistics of the 1870s. Almost all of Saigon's staple trade items, such as rice and cotton, were destined for Asian destinations like Hong Kong and other east coast ports of China, as well as Singapore and, to a lesser extent, the Philippines, with none going to Europe, although they were transported in European and American ships.[16]

The extent of circular mobilities by the locals and Europeans contests the idea that the central highland between South China coast in the east and the Mekong watershed in the west was terra incognita. Most rivers originating west of the central Vietnamese ranges joined the Mekong, while those on the east formed a connection to the various points in the South China Sea through a network that, according to Bennet Bronson's study on coastal Southeast Asia, could be considered a "functional model." In this model, the river integrated its hinterland through a network of tributaries converging around a downstream port, which in turn facilitated connections to the sea. Bronson suggests that the model involves the "control of a drainage basin opening to the sea by a center located at or near the mouth of that basin's major river"[17] (Figure 4.3). Recently, Oscar Salemink has used Bronson's model to explore the specific land-sea connections between the central highlands of Vietnam and the South China Sea.[18]

Bronson's model is generally applicable to all the downstream BISMRY rivers, but the idea that the "interfluvial countryside" beyond the outer edges of the tributaries of the main river was lost in wilderness is not tenable. This was precisely an area where the network of another major river was accessed by local people, opening up larger land-water network. The Mekong networks accordingly spanned Yunnan, Luang Prabang, Moulmein, Bangkok, Phnom Penh, Saigon, the eastern Vietnamese coasts, and China via the Red and Chao Phraya Rivers. In the context of trade networks within a larger

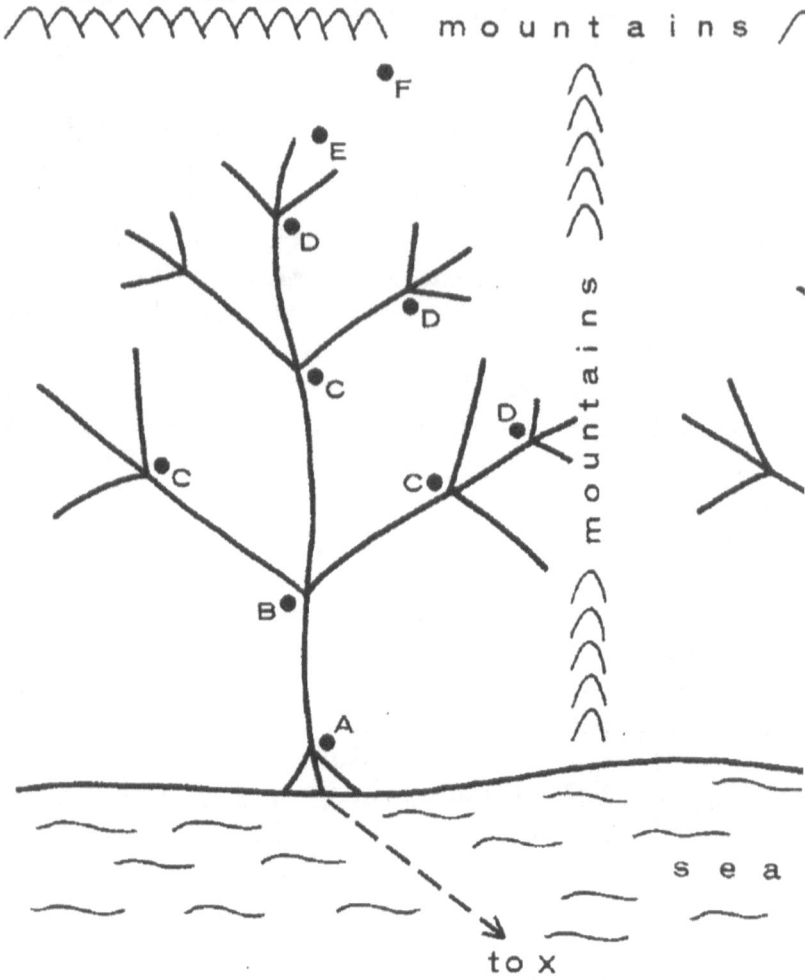

FIGURE 4.3: Bennet Bronson's functional model. Source: Bennet Bronson, "Exchange at the Upstream and Downstream Ends: Notes Toward a Functional Model of the Coastal State in Southeast Asia," in *Economic Exchange and Social Interaction in Southeast Asia: Perspectives from Prehistory, History, and Ethnography*, ed. Karl Hutterer (University of Michigan Press, 1977), 42.

spatial configuration, the Mekong was more like an architect of a massive fluvial network than a mere conduit. For this reason, despite general non-navigability, the Mekong and its trade network remained the apple of discord among the imperial and regional powers. The following section will locate the Mekong within the vortex of these power plays.

II

The French imperial efforts to reach China via the Mekong appeared to Milton Osborne, as they did to French explorers Doudart de Lagrée and Francis Garnier a century earlier, as part of Europe's "universal goal" that harkened back to Marco Polo and stretched into the present when the political route to China was "still uncertain for the Western World."[19] Europe did reach the Mekong Valley, but the empire it sought to build was merely a mimicry of earlier empires shaped by the Mekong, and was hardly translated into a European project. By the turn of the nineteenth century, Thailand and Vietnam had emerged as two ambitious regional powers, with Cambodia and Laos being the bone of contention between them. The mounting conflicts led to a protracted war ending with an agreement in 1847 that confirmed Vietnamese influence over Cambodia and Siamese influence over Laos.[20] In hindsight, this agreement was a recognition of Siamese control of the upper Mekong and Vietnamese control of the lower Mekong. The subsequent French entry into the power play followed in the footsteps of the Nguyễn Dynasty. Vietnamese expansionism was now set on a French footing as both Cambodia and Laos came under French control in the early 1890s, as if the Nguyễn imperial dream of controlling the entire stretch of the Mekong below the southwestern Chinese frontier came true. It is not unreasonable to suggest, as some historians have, that if the Mekong was navigable in its northern course below Yunnan, the French would have been the first to knock out the Konbaung Dynasty in Burma.

After Britain annexed upper Burma in 1885, French ambitions in the region were effectively curtailed, as the British turned their attention to the upper Mekong, viewing it as a strategic counterbalance to growing French influence in Indochina. Thai hostility toward the French, fueled by the British, naturally aligned with British interests, particularly in securing the trade routes to China via Moulmein and Bangkok.[21] These imperial and regional politics came to a head by the early 1890s. The successful annexation of upper Burma emboldened the British to occupy the Shan States and halt French advances on the left bank of the Mekong up to China's borders. The situation grew more complex as Thailand and France locked horns over the demarcation of the Laotian-Vietnamese border.[22] In 1893, France requested Thailand to recognize its rights on the left bank of the Mekong. The British apprehension grew over the prospect of French control extending to the left bank of the upper Mekong, potentially bringing them into direct contact with Burma due to the river's two westward bends.

There was also a nascent British proposal to establish "a neutral intermediate State" between China, Siam, Indo-China and British Burma, centered on the Mekong. Although the "Mekong Buffer State" never materialized, this underscores the complexity of border negotiations in the region. By October 1893, despite British opposition, France asserted its rights over all territories on the eastern bank of the Mekong, from Cambodia to Yunnan's borders, forcing Thailand to relinquish its authority.[23] In 1904, France expanded its holdings to include territories on the right bank, including Champassak.[24] Celebrating the French gains by the turn of the twentieth century, Prince d'Orléans exclaimed: "Once more, greeting to the vast river, over and again purchased to France by the blood of her soldiers, by the lives of her explorers, and by the achievements of diplomacy!"[25]

The apparent gain for the French in the Mekong basin indeed came on the heels of a series of calamitous events, echoing d'Orléans's comments. Early explorer Henri Mouhot had reached Luang Prabang, which he described as a "delicious city," but he passed away soon after heading upstream north of the city. The arrival of the Mekong Mission of Doudart de Lagrée and Francis Garnier marked a milestone in French explorations, but Lagrée also died in a place near Kunming. Garnier managed to reach Dali, but local resistance forced him to abandon his northward journey and return to France via Yangzi. When Garnier returned to the region a few years later, he hoped to connect Yunnan via the Red River rather than Mekong, but he too met a tragic death on a battlefield by the Red River.

In the broader context of French imperial expansion, the territorial and political gains in the Mekong region lasted for only about three decades before the Japanese arrived, and these gains were hardly unmixed. As Milton Osborne notes, even as late as the eve of World War II, the river journey from Saigon to Luang Prabang took longer than the journey from Saigon to Marseilles, suggesting that the Mekong was anything but the "golden route" to China.[26] This dismal scenario of the Mekong from an imperial perspective resulted largely from the frustration over the river's inaccessibility despite the availability of steam technology. The navigability of the Mekong was crucial to the French for strategic reasons, as the British already controlled the Ganga, Brahmaputra, and Irrawaddy basins. The discovery of the Mekong's nonnavigability was, therefore, particularly disappointing. On one occasion, the French had to use the Chao Phraya River to operate gunboats in claiming rights on territories on the Mekong.[27]

The British, on the other hand, never seriously claimed the Mekong, although they occasionally pressed for their rights. They were content with their stake along the five-hundred-mile Chao Phraya basin and the frontier with Siam up to the Mehuok's confluence with the Mekong. British interests in the Mekong Valley were considered less vital because the river could "only be navigated with difficulty, and because France is for some distance athwart the Mekong and commands its mouths."[28]

Amid the French and Thai claims on the Mekong, China remained an active stakeholder in the border talks concerning the river. With its weakening grip over Shanghai and Canton, China sought to secure access to the Mekong, Salween, and Chao Phraya networks as alternative commercial and strategic corridors. However, the Chinese approach to the Mekong south of Yunnan was characterized by what could be termed passive attachment. A British official noted: "Between the Mekong and the Burmese frontier (district to the west and south-west of Simao) there is not one single Chinese official. The whole of that country has no real value in the eyes of the Chinese, and yet, in the fashion thoroughly Celestial, they will strenuously oppose any suggestion made by the British during the next year or so to extend the Burmese frontier a little nearer to the Mekong."[29] To complicate matters, the French themselves urged the Chinese to resist any British claims for territory that "approaches too near the Mekong."[30]

When the Japanese arrived on the scene with their occupation of Indochina in 1940, the contest for political control of the Mekong was in such flux that the deepest channel of the river was taken as the dividing line between French and Thai forces. Both sides were forced to withdraw ten kilometers from the river, creating a "neutral" zone occupied and administered by the Japanese until their defeat.[31]

As World War II came to an end, no imperial power was able to establish absolute authority over the Mekong system for a considerable period, highlighting the contestations and instability in imperial relationships with the river. Previously, Siam had been left as a buffer zone partly because the Mekong was not fully navigable. The river's unnavigability proved to be a force in itself to frustrate imperial plans to establish control of the basin. This audacity of the Mekong survived the ambitions of the British, French, Thai, Chinese, and Japanese Empires, carving the trade routes that these empires were compelled to follow.

The trade routes from larger Mekong river towns, such as Luang Prabang, to Bangkok via the Khorat Plateau were too well established to be fully diverted down the Mekong-Saigon route. French efforts in this direction eventually failed, and realizing the importance of economic connections across the border, they "went to considerable lengths to encourage them."[32] On the river's east bank too, most commerce was in the hands of the Chinese merchants who exported cardamom, stick lac and benzoin, hides and skins, ivory and antlers, among other items, to business houses in Bangkok.[33]

The much-coveted railways also failed to materialize in the Mekong basin. Proposals included a line from Moulmein to Simao with stations at Chiang Mai, Chiang Saen, and Chiang Hung.[34] Hallett had proposed a line from Raheng on the upper Chao Phraya River to Chiang Saen on the Mekong, with a branch line from Chiang Mai, connecting all the major river confluences.[35] Other proposed routes included a line from Kunlon to the Mekong, but China showed little interest, neither endorsing the project nor pursuing its own railway construction in the region.

Each of the empires had a tenuous hold along the Mekong network, despite having access to modern tools and technologies. Over the nineteenth century, the Thai and Vietnamese polities became prominent, only to be displaced by the French, whose dominance was short lived and ended by the Japanese. Throughout these changes, the Mekong itself remained the steadiest figure in the current of events. These processes led to a chaotic period of border making and unmaking with serious regional implications as reflected in much of Southeast Asian studies. Amid shifting transimperial power struggles, the only force that looked aligned with the Mekong's footprints were the riparian ethnic communities, to whom we now turn.

III

Despite ongoing rivalries among imperial powers, the imperative of maintaining trade flows along the Mekong network ensured that long-distance mobility remained largely uninterrupted. However, in certain areas of imperial encroachment, ethnic communities resisted, striving to reclaim alienated trade corridors and assert their presence. The Akha people, for example, originally from a region spanning Sichuan and Yunnan, migrated southward and settled along the Mekong's banks and its tributaries. Over time, they expanded their mobility network, extending as far south as the Khorat Plateau.

They maintained a circulatory movement, always returning to their home region, believing that a journey "has to end where it had started."[36]

With the French advance, the Akha community found themselves divided and restricted in mobility due to the demarcation of the Mekong as the border between Siam and Indo-China. The French took over the left bank of the Mekong while Siam controlled the right bank in Lao territory of the Khorat Plateau. Furthermore, the French and the Siamese imposed new taxes on territories along both banks after the 1893 agreement, and the opium and slave trade were discontinued. By the turn of the century, these disruptions led some Buddhist monks to rebel, fueled by millenarian expectation that the next Buddha would unify the peasants from both sides of the Mekong. These rebellions continued in some forms until the arrival of the Japanese, although the Siamese had already suppressed some of them.[37] Like the Abors and Kachins in the Brahmaputra and Irrawaddy headwaters, the Akhas resisted the intervention of the Siamese and the French as their access to the Mekong network was disrupted.

Resistance against the Chinese also emerged when local ethnic groups' access to the Mekong trade network was threatened, as control over these routes was vital for their economic and social survival. The Loutzu's relationship with the Chinese often depended on the question of such accessibility. Yingpan, a town in northwestern Yunnan on the Mekong, was populated by the Loutzu people. It stood on the cotton route from Chao Phraya and was connected to the salt trade from the nine salt mines on the east bank of the Mekong in Yuntung Subprefecture and other neighboring areas. Despite several attempts, the *lo t'ukuan* (Chinese hereditary officials) couldn't coerce the Loutzu to pay tribute, resulting in frequent skirmishes and a state of "wild independence." Nevertheless, Chinese traders traveled through their territories, and the Loutzu engaged in trade, exchanging gold, drugs, beeswax, skins, hemp, and lacquer against salt, cotton, and tobacco, among other items.[38] In the lower Mekong region, as David Biggs has shown, the ecological fluidity enabled local groups to evade colonial encroachment and offer active resistance.[39]

In conditions where external interventions were not strongly felt, ethnic relations resembled those of the Shan traders in the Irrawaddy and Salween basins. In the lower Mekong Delta, the Cham Muslim community exemplified strong links between ethnic identities and the availability of fresh water for agriculture and commerce, as Philip Taylor's extensive ethnographic works have shown.[40] Tuned to the idea of "looking at the river, thinking of the sea," the

Chams maintained trade relations with the central Vietnam highlands, where their Hindu ethnic cousins had lived for centuries. This formed an important trade network between Phnom Penh, Saigon, and the central highlands.[41]

Luong observes how historical connections between the two regions persist through the seasonal migrations of agricultural laborers and retail vendors, facilitating trade in coffee, cashew, and peppercorn from the central highlands to the Mekong Delta.[42] The Hmong, usually concentrated in the upper Red River basin, found themselves along the trade routes between the Mekong and Chao Phraya systems across Chiang Rai and Chiang Mai during the nineteenth and twentieth centuries. Hmong settlements around Nan River basin in Chiang Rai can be understood in this commercially strategic sense, as trade relations between the Mekong Valley and that region were strong.[43]

The Yao people continued to maintain a significant presence across the region. They were among the latest ethnic groups to settle in northern Thailand from Laos during the second half of the nineteenth century. Their main groups took three different river routes: one group came from northwest Laos around Luang Prabang and settled in Nan; another also came from northwest Laos, crossed the Mekong, and settled in Chiang Rai; the last group arrived from upper Burma and settled on the upper reaches of the Mekong, intermittently among the Chinese Shans, Meo, Akha, and Karens.[44] By settling in areas that connected the Mekong Valley with Bangkok and Moulmein via the Chao Phraya and Salween, these people demonstrated a preference for the riverine commercial trail that transcended imagined divisions between upstream and downstream or mountain and plainland.

If the Chams, Hmongs, and Yaos made translocal mobility a part of their lives, others opted for what could be termed rooted mobility, as noted earlier. Some confined their movements to the tributaries and major nodal points, which often happened to be confluences (locally known as *paak*) like Luang Prabang, Bhamo, or Sirajganj. The Lamets in the northwest corner of Laos, living between the Ngao and Nam Tha Rivers near the current border with Thailand and Burma, showed good examples of rooted mobility.

As Robbins Burling notes, immediately after the harvest, the Lamets went to the markets along the Mekong to sell rice, forest products, deer skins, honey and beeswax, baskets, a type of rice brandy, and fermented tea leaves. Other items included horns and dried gallbladders of deer, which were sold to the Chinese who used them for medicinal purposes. In return, the Lamets obtained iron tools, cloth, pottery, and sometimes secondhand felt hats, glass bottles, cigar lighters, and cotton blankets. They were diligently "dependent on markets and through these markets, have long had access to the products and ideas of the world beyond their borders."[45]

In another detailed study, Izikowitz notes how the Lamet ironsmiths moved along the Mekong and its tributaries to trade with the mountain people and Laotian merchants, especially following the harvest when they needed new knives to reclaim land for swiddens. In the preharvest incantation dedicated to the paddy spirit, among many other deities, the Lamets wished to "have copper billets from Nam Tha, . . . have coins from Mekong, have pleasant sounding bronze drums, have good wives, have big sliver ornaments, and buffalo with long horns." In other words, they wished for "enough rice and money to enable them to buy wares from merchants that come up the Nam Tha and Mekong Rivers."[46]

The Lao were more widely distributed as craftsmen and traders compared to the Lamets. During the paddy harvesting season, Lao merchants gathered at a *paak*, or certain other places on the Mekong, where the Lamets exchanged rice, which the Lao distributed in Luang Prabang and villages along the river. The Lao also paddled their canoes up the tributaries of the Mekong to sell their products to mountain dwellers. The relationship that developed between the two people was mutually beneficial. As Izikowitz notes: "Some of the Lao settled farther up the tributaries in order to be nearer their clients in the mountains. It is also possible that they wanted to compete with those living near the mouths of the rivers." Thus the Lao settled where it suited them best commercially—for example, where "a path met a river, a fork, or a place that mountain people could easily reach with their bamboo rafts. Agriculture took second place." Izikowitz further notes, "As a rule the Lao have a habit of remaining near the mouths of the tributaries of the Mekong, where they buy up the products which the Khmu, the Lamet, and other mountain tribes deliver."[47]

The Lu people demonstrated the advantage of the mountains in this flux of mobility. They used the path leading to Mokala Panghay from the south as it connected the ancient caravan trail along the Nam Ngao with the river route of the Nam Tha. In their prayers, they summoned the spirit of mountains, horizons, rivers, and valleys, expecting the rice souls from the territories around the Nam Tha and Nam Ngau rivers to "climb over the mountains and unite in their own swidden." Following the harvest, the villagers put up communal efforts to keep roads to the river clear in order to facilitate trade with the Lao people, who apparently embodied some of the agencies that the Lu people invoked in their prayers.[48]

Within this context of multimodal mobilities, as Andrew Walker points out, Wittfogel and Wolf's ideas of tributary forms of power relations and static forms of subsistence were not the dominant features of the region. Although evidence is fragmentary, precolonial social formations of northwestern Laos were characterized by "more multifaceted internal and external trading linkages than the 'tributary model.'" This suggests that precolonial states of northern Laos actively and enthusiastically supplemented their tributary income with revenue from trade and that, in some cases, tributary and mercantile institutions were "closely interlinked," and for Luang Prabang this was the "basis of its economic survival."[49] Walker further notes that the commercial acumen in the Mekong basin was connected to a hybrid "local landscape of power" that derived "not from isolation, peripherality or marginality but from connectedness."[50]

IV

The tripartite political and commercial entanglements among external colonial powers, regional forces, and local ethnic groups over the Mekong set the stage for an affective commoning process within the river network. The delta, stretching westward to Siam, along coastal water routes to the Malay Peninsula, and into the eastern montane region, fostered a dynamic riverine-maritime nexus. This nexus created what Nola Cook calls "Water World," a condition beyond national borders, characterized by "kaleidoscope" of local and migrant people.[51] With a focus on the Cham, Khmer, Kinh, and Chinese actors, Philip Taylor outlines an impressive picture of the cosmopolitan nature of the Mekong Delta.[52] The plurality of the lower Mekong water world was a reflection not merely of its exposure to maritime cultures but also its connection to the Mekong network, extending to Yunnan, Sichuan, and Tibet. The previous chapter on the Salween noted Chiang Mai's cosmopolitan interface at the fluvial nodes. A little more in-depth discussion is warranted here about its specific connection to the Mekong.

A major contributor to the cosmopolitan space at Chiang Mai was the Hui Muslim caravanners from Yunnan, many of whom settled in the Ping River Valley where Uttaradit was the starting point for the barge navigation on Chao Phraya River to Bangkok. Following the rebellion, a significant number of Hui settled in Chiang Mai, bringing with them a commercial network that spanned Yunnan, Tibet, Laos, and northern Thailand.[53] Soonthornpasuch's story of a postrebellion Hui settler, Ch'un Chowng-lin, illustrates this cosmopolitan flux. Ch'un's trade traversed Kunming through Kengtung in the Shan States into Mae Sai, Lampang, Tak, Lamphun, and Chiang Mai. He married a local Thai woman, and before settling in Chiang Mai in 1915, the couple exported goods from Mae Cham, a district to the northwest of Chiang Mai province in the Salween basin, on land granted by the Thai king. The site where Ch'un first settled was initially a site for Yunnanese traders to unload their goods and rest their pack animals. For his business achievements, initiatives, and charitable activities, Ch'un Chowng-lin was granted knighthood.[54] Throughout his career, he cemented the bonds between the people of the Salween, Chao Phraya, and Mekong Valleys with Chiang Mai at the center.

Further up the Mekong at Luang Prabang, the lively cosmopolitan vibe was replicated in an environment that was "certainly the largest and most crowded of any in the north, with the exception of Chiangmai," as Forbes noted.[55] It was connected to Chiang Mai via the Nan River (with Uttaradit being the point of navigation) in the south, and to Yunnan and surrounding areas via the Mekong network. Lamington offers a detailed sketch of the external entanglements of the surrounding areas, as he encountered various ethnic groups. Some items sold in the neighborhood included goods from Manchester, traded by Yunnan muleteers.[56] To Lamington, Muang La, in the vicinity of Sibsangpanna near the Nam Pak River, resembled a "Swiss village" with its little streets. In the Muang mining area, he found "people so busy . . . cheerful and extremely pleasant to talk to." In the surrounding villages, especially near the Nam Noi and Nam Ma Rivers, there was an abundance of resources. He met with women from whom he requested samples of homespun petticoats, and was offered a "great variety to choose from," while the women themselves wore clothes of imported red and white silk or velvet.[57]

At Xiaguan, a commercial town of West Yunnan south of the Dali Valley located more than a thousand kilometers from Saigon, the vibrancy of the lower Mekong trade was hardly diminished. The town was located on both sides of a river that flowed from the southwest corner of Erhai or Dali Lake and joined Yangpi River, another tributary of the Mekong, augmenting the town's commercial agility. George Litton noted that rents in the town were higher than in Dali, and of the twelve thousand population, "nearly all" were traders, reflecting on its nodality as a regional hub amid extensive translocal connections. In addition to opium production, the town was a vibrant hub for various traders, including dealers of foreign goods from Tengyueh, cloth sellers, opium buyers from Linan (south Yunnan), and merchants from Canton. It also hosted peddlers from Sichuan and traders from Lijiang, Tibet, who exchanged ponies, mules, woolen cloth, drugs, and musk for tea, salt, sugar, and cotton.[58]

The Mekong's connections to Canton, Sichuan, upper Burma, and the Chao Phraya system were complemented by its links to Tibet, Assam, and the eastern Himalayan mountain fringes, extending all the way to the Chittagong Hill tracts in eastern Bengal. This network meant that the Mekong's commercial and cultural flows were as much vertical as they were horizontal. In 1889 Kuhn stated that the Khasi language spoken in the Meghalaya and Sylhet region had close affinities with the Moo-Anoam language spoken in parts of the upper Mekong. He noted, "From both sides, from language and words and

FIGURE 4.4: "Bridge over branch of many-mouthed Mekong, and native craft, Saigon, Cochin-China," 1915. Source: The Library of Congress. https://www.loc.gov/resource/stereo.1s40273/

Passing a Mule over the Mekong at Tsedjrong.

FIGURE 4.5: "Passing a Mule Over the Mekong." Source: Prince Henri d'Orléans. *From Tonkin to India by the Sources of the Irawadi, January '95–January '96*, trans. Hamley Bent (Methuen & Co, 1898), 219.

sentence formation, the Khasi language has close affinity with Palaung-Wa (Palaungic)/Austroasiatic spoken by people living beside the Mekong river. "[59]

The Mekong's nodal points at its confluences with tributaries were often the sites of bazaars or towns. These confluences were traditionally known as *paak*, a word that conveys a similar meaning in the Bengal Delta: "muddy whirlpool." These locations have long been popular sites for settlements, and numerous villages are identified by this word: Paktha, Pakkhop, Pakngenuy, and Pakou.⁶⁰ These *paaks*, therefore, can be seen as a material basis of cultural and spiritual flows. Lamington noted such a Mekong confluence with its eastbound tributary, the Nam Ma, at Chieng Lap, where "some tapers had been burnt to propitiate the devils, and a sacrifice of rice, nuts, and a rupee offered up on the rocks."⁶¹ Lisus, Shans, Karens, and Lahu, among other groups, all venerated spirits that together entangled mountain, river, agrarian, and commercial spaces that transcended physical boundaries.⁶²

The Mekong's inner spaces were foreclosed to any large-scale imperial project of exploitation but remained open to messy entanglements of people and products through transregional trade relations shaped by an intimate relationship with the riverscape (see, for example, Figures 4.4, 4.5, and 4.6). This intimacy was reflected in the deep knowledge that the people displayed, for example, in mapmaking. Hallett encountered some local chiefs who made a map of the country with matches and bamboo strips on the floor, showing the position of the sources of the Meh Ping, Meh Teng, Meh Hang, Meh Pai, Meh Nium, Meh Pam, and other streams.⁶³ Another example of local intimacy

FIGURE 4.6: "Launching a Boat at Luang-Prabang, on the Mekong." Source: Archibald Ross Colquhoun, *Amongst the Shans* (Field & Tuer, 1885), 287. The Digitized Collections of the Staatsbibliothek zu Berlin.

FIGURE 4.7: "A Boat ascending a rapid on the Mekong." Source: Archibald Ross Colquhoun, *Amongst the Shans* (Field & Tuer, 1885), 10. The Digitized Collections of the Staatsbibliothek zu Berlin.

with the Mekong was the intuitive and practical skills required to navigate the difficult parts of the river, which were as challenging as traveling on the ocean, where the experience was no less frightening than a ship being tossed on the tip of giant waves. But seasoned Mekong canoe boatmen could manage such situations with "lightning speed"[64] (see Figure 4.7). For those who came from outside this riparian wilderness but longed to engage it as it was, the challenge and gratification of the encounter were shared. As Kingdon Ward described his encounter with the upper Mekong: "The traveller, buffeted and bruised by storm and mountain, cherishes most of the foe worthy of his steel. Nevertheless, there was a strange fascination about its olive-green water in winter, its boiling red floods in summer, and the ever-lasting thunder of its rapids. And its peaceful little villages."[65]

Human mobility, as reflected in the accounts of the surrounding landscape and in the intimate encounter with the fluvial "manifestedness" of nature, was also linked to the common quest for peaceful dwelling, which Kingdon Ward experienced in a "peaceful little village" on the Mekong (Figure 4.8). This is the peace of dwelling in nature with all its charms and challenges, which the Mekong offered in abundance—a sentiment that echoes Heidegger,

108 Chapter Four

CAMBODIAN HUT AT PEMPTIELAN, ON THE MEKON.

FIGURE 4.8: Dwelling and moving by the Mekong. Source: M. Henri Mouhot, *Travels in the Central Part of the Indo-China*, vol. I (John Murray, 1864), 232.

a contemporary of Kingdon Ward: "To dwell, to be set at peace, means to remain at peace within the free sphere that safeguards each thing in its nature. The fundamental character of dwelling is this sparing and preserving. It pervades dwelling in its whole range. That range reveals itself to us as soon as we reflect that human being consists in dwelling and, indeed, dwelling in the sense of the stay of mortals on the earth."[66]

The Mekong posed the same challenge to the French Empire as the Salween did to the British in terms of a lack of unhindered commercial navigation facilities from the ocean to Yunnan. Despite having visible military and political clout over the Mekong, the French could retain their hold over various parts of the river only by allowing the continuity of precolonial mobility of indigenous agrarian and commercial forces. In the French dealing with Champa, Tai, Lao, and Chinese people and other ethnic groups, as well as the Japanese dealing with them during the occupation, a wide range of accommodations was made for various communities and forces along the Mekong. In the absence of monopoly by any single power over the river, commercial and strategic options had to come to terms with the geomorphological challenges presented by the Mekong system. As a result, the river continued to be an open space, touching and shaping the everyday life of different cultural, ethnic, and commercial groups spanning the Mekong Delta and Yunnan. Its unnavigability induced enhanced interactions with neighboring river networks such as the Irrawaddy, Salween, Chao Phraya, and Red. As we shall see in the next chapter, the Red complemented the Mekong's transregional agility as it spanned the Gulf of Tonkin and Yunnan, and parts of other BISMRY basins.

FIVE

THE ORIGINAL RIVER
The Red Beyond the Nation and the Empire

WITH ITS FULLY NAVIGABLE parts within the country's current border, the Red River is the nerve line of Vietnam.[1] As the lifeblood of the millennial capital city, the epicenter of Minh civilization, and the provider of a dynamic agrarian and commercial economy, the Red represents a reasonable recipe for a national imagination. A fondly remembered period of Vietnamese anticolonial resistance along its basin adds extra appeal to these collective aspirations. Although the Mekong remains a significant national site, only 8 percent of its basin lies within Vietnam, and the honor of being the "cradle of the Vietnamese nation" has been reserved for the Red.[2]

This national narrative of the Red has recently been revisited in the literature that highlights long-term histories of transregional mobility along its corridor toward the South China Sea on the one hand, and the BISMRY river networks on the other.[3] Of all the rivers meeting the waters of the Indo-Pacific, the Red reaches the border of China quickest. In terms of transport facilities from the last navigation point to central Yunnan, as Nanny Kim notes, the Red River route was more accessible than any other neighboring rivers.[4] These land-river-ocean mobility networks within a relatively shorter distance led to more trade between China and the Red Valley than between China and the rest of Vietnam, both in historical and contemporary times.

In keeping with the findings on other BISMRY rivers, this chapter explores the Red's transregional itinerary and attendant political, economic, and ethnic relations that fostered shared engagements. The quest for imperial dominance over the Red River network was intense, and so were the contestations against it. Between the terrestrial space that the French sought to exploit for trade and commerce and the local pressure groups that resisted such advances, the Red network emerged as a moving mediator. The flow of humans, nonhumans, and commodities straddling these political contingencies collectively created a third space around the river system, whose agency transcended both the nation and the empire.

I

In its nearly 1,100-kilometer course between the Gulf of Tonkin and the Dali region where it originated, the Red River's major nodes included Haiphong, Hanoi, Yen Bai, Lao Cai, and Manhao. Haiphong was not situated on the mouth of the main flow of the Red itself but on the Cam, which in its turn was connected to the Red through an intricate river network. European explorers naturally saw Haiphong as a launchpad for overland trade spanning the West River and the territories south of the Yangzi in a view to connecting Canton.[5]

However, during the nineteenth century, much of the West river-bound traffic from Yunnan was diverted either to the Red or the Yangzi. Consequently, Haiphong's primary connections with Yunnan developed via Hanoi, leaving Guanxi "locked in the lee of international trade."[6] Between 1882 and 1891, the tonnage of export and import in Yunnan via the Red increased exponentially from 886 to 5,234, with Hong Kong and Tonkin accounting for approximately 85 percent and 15 percent of the trade, respectively.[7] The Red River became so central in these transregional flows that Australian traveler Morrison asserted that no artificial means could alter the "natural highway" of central and southern Yunnan with Tonkin.[8]

Hanoi symbolized the beginning of the Red River Delta following the convergence of the river's two main tributaries: Black (Da) and the Lo river. Hanoi's location below this confluence resembled that of Dhaka below the Brahmaputra-Ganga confluence, Pagan below the Chindwin-Irrawaddy confluence, and Phnom Penh around the Mekong–Tonle Sap confluence. Each of these rivers was a collection of a "hefty number of streams." The designation of Hanoi as the "Navel of the Dragon" (*Long Do*) was indicative not only of its strategic location in the delta but also its importance as a hub to various water bodies, including streams, lakes, and the three rivers, which functioned as twelve "water gates" to the city center and its markets.[9] This fluvial multinodality informed the local proverb that imagined Hanoi as the head, and Ninh

Binh the neck, from where many small connected rivers made "navigation everywhere possible for junks."[10]

From Hanoi upward, the flow of people and commodities was centered on Lao Cai where Chinese commercial activities predominated. In 1908, for example, Chinese Customs in Hekou Yao (across the Red River from Lao Cai) recorded the clearance of 10,000 boats carrying 30,000 to 40,000 tons of cargo.[11] Typically, commodities flowing between Hanoi and Lao Cai were transshipped at Yen Bai to smaller boats and junks heading to Lao Cai, and to larger boats and ships destined for Hanoi and Haiphong.[12] Trade activities became even more significant further upstream in Manhao, a river port on the Red within the Chinese border. At six hundred feet above sea level, Manhao was the last major navigable point on the Red, accessible only to junks and smaller boats due to about fifteen strong rapids above Lao Cai. Despite being described as "small," these boats could be as long as 50 feet and 10 feet wide capable of carrying a load up to 10,000 pounds.

Although the Red River ceased to be navigable at Manhao, its extended life and network grew wider as it became a crossroads for cargo to different inland destinations. Ferdinand von Richthofen, a Berlin (Humboldt) University professor and coiner of the term "Silk Route," provided eleven arguments in favor of the Red River as a passage to China. He pointed out that compared to the distance from the last navigable points of other BISMRY rivers to Kunming, Manhao was the nearest, requiring 12-day journey. In comparison, Bhamo on the Irrawaddy was 28 days away, Yibin on the Yangzi 24 days, and Kiang Hung on the Mekong 24 days.[13] At Manhao, Prince Henri d'Orléans observed "processions of draught oxen" during his 1894 travel, and one morning, he saw a caravan of at least 130 mules, carrying loads of tins among other products in exchange for linen, yarn, and tobacco. He discovered that junks operating at Manhao often had a tonnage of 5,886, with tin being a major item shipped to Hanoi and then to Hong Kong. Manhao was so busy that d'Orléans couldn't hire mules for his travels, as all were contracted to the merchants of Mengzi.[14]

Next to Manhao, Mengzi was another land-based nodal point. Although the 37-mile trip between Manhao and Mengzi involved a steep rise in elevation from 510 to 6,150 feet, the trade flow remained significant. All Chinese produce coming down to Tonkin from Yunnan and Guanxi enjoyed tariff-free transshipment at Mengzi, and trade continued to grow throughout the late nineteenth century. Between 1890 and 1895, total trade at Mengzi increased from 1,104,007 taels to 2,842,319 taels.[15]

A considerable portion of this trade involved the export of tin and copper and the import of various products, including salt from Hong Kong via Manhao.[16] Raw cotton and cotton threads of Indian and Japanese origin (more than 30 percent), prepared tobacco from Canton, coffin woods from Tonkin, needles and metal buttons, and miscellaneous items from Britain and France took the Red River route to reach Yunnan via Mengzi.[17] Indian yarn was considered cheaper than the Manchester product and rapidly replaced the latter in western China.[18] On the export side, Mengzi sent out mainly tin slabs, silk, opium, lac, and miscellaneous items for the Red River route. In 1883, Colquhoun reported about 20 million francs of trade, with imports slightly less than exports on this river route.[19]

A British official report from 1898 suggested that nine-tenths of the goods imported to Yunnan followed the Red River route.[20] After meeting the needs of Yunnan, some products found their way to regions beyond the province, though full statistics on this may have remained unrecorded. From Mengzi, numerous routes branched toward the Yangzi Valley, including destinations such as Guizhou, Changsha, Chengdu, Hubei, and parts of Sichuan, which accounted for about 14 percent of imports at Mengzi. The Red-Mengzi route also extended to Hsu-chou Fu in Sichuan at the confluence of the Jinsha and Min Rivers.[21]

Other destinations of these commodities entering Yunnan included the bazaars of the Irrawaddy Valley via Bhamo and parts of the Salween.[22] In 1870, Augustus Margary, a British diplomat and explorer, bought silk and copper in the Moulmein bazaar from a group of twenty-eight men who were inhabitants of Tonkin, and had traveled to Moulmein in ninety days. This finding was later confirmed by T. T. Cooper, who recognized the type of silk specimens brought by these "hardy" traders.[23] From the Irrawaddy network, a major product that reached the Red Valley was Bombay cotton, which formed the bulk of commodities that came on the back of at least a thousand mules from Bhamo, even during the prohibitive rainy season.[24] Another significant product connecting the Irrawaddy with the Red Valley was jade from upper Burma, a prominent feature of the traffic between Yunnan and Canton via Dali, Mengzi, and Manhao.[25] Additional intervalley destinations from Manhao and Mengzi included Sadiya on the Brahmaputra, connected via Dali and eastern Tibetan territories.[26]

The Black River enabled another cluster of routes around the Mekong Valley in the northwest direction via Simao and Pu'er, which was crisscrossed

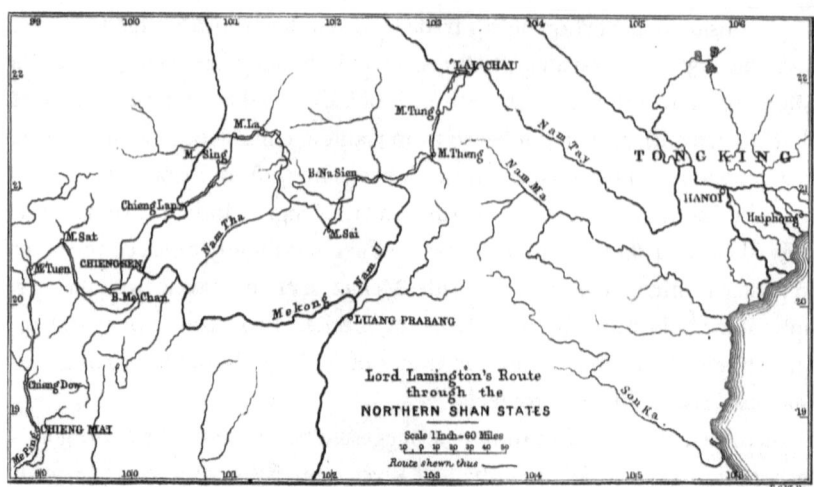

FIGURE 5.1: The Red's connection to the Mekong-Salween networks. Source: Lord Lamington, "Journey Through the Trans-Salwin Shan States to Tong-King," *Proceedings of the Royal Geographical Society and Monthly Record of Geography*, 13, no. 12 (December 1891).

by the Babian, Manlao and Pu'er Rivers. In the nearly 350 miles between Manhao and Simao, located between the Black and the Mekong Rivers, Simao served as a crossroads through which at least four routes reached out to different transit points on the Mekong. At the same time, each "insignificant village is connected with the one next to it" via multiple road networks.²⁷ Additionally, six forking roads extended to Dali in the north, the Shan States in the southwest toward Burma, and Luang Prabang in Laos. In other words, this area formed a grand Red-Black-Mekong network (Figure 5.1).

Of the 40,000 *piculs* of tea produced in the neighborhood of Simao, a significant amount was processed and packed for export throughout China, including the royal court and "everywhere from Bhamo to Shanghai," as Fred Carey noted. Simao also served as an entrepôt for cotton transported from Bhamo and the Shan States to Kunming and Sichuan across the Mekong. The amount of cotton processed at Simao for Yunnan was about 16,000 *piculs* annually.²⁸ Simao, according to d'Orléans, had a population of about 10,000, but the "floating population" of caravanners and muleteers living in the inns on the town's outskirts was "considerable."²⁹ These existing intervalley links prompted some observers to envision practical means of communication between the basins of the Mekong and the Gulf of Tonkin, the bare possibility of which "lend additional fascination to the study of this region."³⁰

Overall, as Jean Michaud noted, the Red River remained an active site of human and commercial flows despite topographical challenges. He summarized the nodality of the Red in its regional and transregional setting in these words:

> Despite scanty infrastructure, the basic elements of trade networks did exist in, across, and beyond these mountains, providing upland markets with goods not available locally such as salt, metals, gunpowder and utilitarian industrial goods that were essentially lowland commodities. These networks operated mostly through the main river systems, sometimes across the Chinese border. More often than not, Chinese merchants were involved, and the waterways of the northern mountain ranges connected trade between Yunnan, Guangxi and the Red River Delta.[31]

Michaud's observations highlight why the Red River network and the extensive commercial mobility it sustained attracted renewed attention from imperial forces involving the French, Chinese, and Vietnamese. The following discussion explores how these dynamics unfolded between the French and other imperial contenders eyeing the Red River.

II

The unification of Vietnam in 1802 under Gia Long (Nguyễn Phúc Ánh) occurred in the looming shadow of the French Empire. By the 1850s, the Ngyuen Empire was in retreat, and the French asserted themselves amid this shift in power in northern and southern Indochina. When the Mekong expedition team led by Francis Garnier concluded that the river was unsuitable for through commercial navigation from Saigon to Yunnan, the Red River's potential was explored with renewed enthusiasm, being imagined as "riche de promesses."[32]

These promises were reflected in the fact that, since 1885, the import trade to Yunnan had "almost entirely shifted" from the West River to the Red River route. This commercial "revolution" was attributed to the French effort to convince the Chinese government to allow transit passes.[33] A French presence on the Red also opened gateways to the upper Yangzi, which was considered an "unquestionable success," especially against the backdrop of navigational failure on the Mekong.[34] Around the same time, clandestine follow-up missions were planned to divert Sichuan trade from its "natural channel" of the Yangzi to the Red. The French "forward policy" envisioned tapping trade to the Red from Dali, Tibet, Guizhou, and Guangxi.[35] French adventurer Jean Dupuis looked further to the west as he followed Yunnan's trade links with the Mekong Valley in Laos and the upper Irrawaddy Valley.[36] Dupuis's supply of arms to the Chinese Mandarins against the Hui rebels in Yunnan and Hmongs in Guizhou around 1870 aligned with their plan for trade expansion. This move aimed to secure the Red's commercial passage spanning the Tonkin Delta and Yunnan with the blessing of Qing China.[37]

The French success in securing trade flows on the Red River was partially aided by the Christian missionaries. Michaud shows that the preaching activities of the French Catholic missions in the upper Tonkin region mostly took place in the river valleys and confluences where the vicariate spanned "southeastern border of Yunnan, the right bank of the Lô and (downstream from the junction of the Hong Rivers to the east), the western limit of the Da River basins on the west, and Hanoi province to the south," indicating that all French missionary settlements were located within the Red-Black River network.[38]

By 1910, after fifteen years of activity, the upper Tonkin vicariate included a total of 115 Christian settlements, nearly all located along the main rivers and connected land routes.[39] Michaud argues that the missionaries were a bridge between the Red-Black profitable trade networks and the French imperial network of explorers, merchants, and military.[40] At Mengzi, for example, Roman Catholic Chinese controlled a "fair amount of trade" in items such as clocks, watches, lamps, oil, glass crockery, dyes, ironmongery, magnets, musical instruments parts, wines, spirits, cigars, umbrellas, and fancy items.[41]

Despite the glimpses of these deepening shadows of the French Empire in Indochina, it had negligible influence around the Red-Black basin. One reason was the inaccessibility and unnavigability of the river networks between the upper Mekong and upper Red systems. As Lagrée and Garnier noted, while traveling between P'uerh (Pu'er) and Yuan-Chiang, it was difficult to ascertain which stream was flowing into which river.[42] From its source near Dali to Manhao, the Red was completely unnavigable. The segment between Manhao and Lao Cai was characterized by steep descent where only light draught boats could move, mainly during the high water season from June to December.[43] Further down, from Lao Cai to Kouence, navigation was "impeded by sand-banks, rocks, and 15 rapids, making it impracticable from November till March except for boats of 4 tons."[44] In the delta region, the Red was navigable for larger river steamers at all seasons up to Yenbay via Vietry, but sand bars limited the draft of vessels to eighteen feet, and beyond Haiphong six feet was the "absolute limit of draught for vessels proceeding to Hanoi during six months."45

At the Red River's mouth, unlike in the Bengal, Irrawaddy, or Yangzi Deltas, most parts were punctuated with bars, making it worse than the Salween mouths, although its territorial extent was over 4,000 square miles, intersected by canals and creeks, and endowed with a coastline of 75 miles. In the seven major mouths of the river, navigability ranged from 18 inches to 9 feet of water. The relative nonnavigability due to the bars was compensated by the Canal des Bamboos, which connected Haiphong with Hanoi via Hung Yen, which, in turn, was connected to the Red with Tai-Binh.46

All these factors meant that the French had to maintain the Red River for continued navigation and organization of towing services, which were primarily possible with the collaboration of the locals.47 Locally built traditional boats were more in high demand, embedding local economy and employment. Gunboats found little advantage in the deltaic water or upstream. In other words, the Red River offered resistance to imperial advance at the very entrance through bars and tidal complexities—even more so than the Salween and the Mekong—which Newton observed in *Principia* more than a century earlier.48

The physiographical challenges the French faced on the Red River were compounded by the Qing imperial resistance both before and after the 1885 Sino-French treaty. Although this treaty appeared to mark the culmination of French success in securing the Red River route, it ultimately favored China as it led to the withdrawal of the French naval blockade of Yangzi. In the long term, the French could hardly benefit exclusively from the Red. In the lower delta region, French flag carriers were few and far between, representing only 5 percent of the tonnage in the 1880s, compared to 35 percent by the English, 23.4 percent by the Chinese and Cantonese, 11 percent by the Germans, and 5.5 percent by the Dutch.49

Moreover, the Chinese threat hardly diminished after the treaty. Dupuis himself noted that a Chinese military official, Colonel Tsai, had told him that the province of Guangxi wanted to take control of the Red Delta to "get rid of this troublesome [Vietnamese] neighbour," as he promised support for the French upward advance along the Red.50 China's subtle strategic threat was all too obvious to the French and it was felt not only in the north but also in the deltaic south. No wonder, to defend Haiphong and Hongai from a possible Chinese attack, the French Chamber for Colonial Defenses allocated 19 million francs.51

The only force that could have secured exclusive French domination of the Red-Black basin was the railway, but that tool came late into the colonial era and secured less gain than expected. In 1898, the Chinese granted the French the right to construct a railway to Kunming—a right that the French had dreamed of ever since their arrival in Indochina. Dupuis had already suggested that if he had received a French arsenal and a railway line linking the Red River with Yunnan, the English would not "recover from that one."[52] The French railway to Kunming was opened in 1910, and within a short period, the railway indeed diverted a significant amount of trade away from the Red. In 1908, for example, the Chinese customs office in Hokow (near Lao Cai) recorded the clearance of 10,000 boats and a total of 30,000 to 40,000 tons, but by 1910 the number was reduced to 2,400 boats and 2,800 tons.

The railway-era decline of water transport, however, was not an accurate reflection of the flow of boats and commodities outside official records, which remained substantial. Tsung-fei suggests that larger boats were replaced by smaller and local boats, which may not have been counted in the official records.[53] The freight charge for the railway was also higher than that for boats, which favored the continuing presence of the latter. Besides, the railways were confined to one side of the Red River banks with limited tonnage, whereas commercial flows on the Red itself were widespread along its tributaries and the roads connected to the river by mules. After twelve years in the making and costing about 200 million francs, the railway, it appears, "never fulfilled the expectation."[54] With the Sino-Japanese War and the relocation of the Chinese capital to Chongqing, the Red River route, along with Burma-Yunnan routes, was significantly revived and "more than regained their lost importance." Tsung-fei also notes that within just decades, World War II had "put the clock back for the Red," where the Yunnan-Indochina railway had to be "dismantled in favour of the time-honoured water route and pack animals."[55] Mules, in particular, like in other upper BISMRY river-land spaces, were both a transporter and partner in mobility (see Figure 5.2).

The imperial powers operating in the Red River basin were also hindered by a lack of immunity from epidemics like malaria, which was rampant in the upper Red basin during the rainy season. This was partly due to the construction of French railways, which exacerbated the already high level of malaria. Particularly below Mengzi, where the railway was constructed across the valley of a tributary of the Red River called Namti, "deadly malaria due to

Embarkation of Mules at Notcha Tian-pi.

FIGURE 5.2: A mule boarding a boat on the Red. Source: Prince Henri d'Orléans, *From Tonkin to India by the Sources of the Irawadi, January '95–January '96* (Methuen & Co, 1898), 96.

what Chinese call Chang-ch'I or poisonous air" infested all the descents from the Yunnan plateau.[56]

Because of this, Manhao was lively only until sundown, as Chinese traders and Europeans left the place before evening for fear of mosquitos. Only a high bribe, as Little observed, would induce a Yunnanese to pass a night there or even to descend within a day's journey of the Red River. Little also observed that the Chinese, except for the locals who were born and lived there, seemed "far more susceptible to malarial influences than even Europeans."[57] In Xishuangbanna, half of the region was so malarious that the Chinese administration or military had little sway, extending no more than a few miles to the west or south of Simao. However, unlike the French or the Chinese, the Shans or the Lolos, for example, were not affected by malaria.[58]

During their advance in Indo-China, the French were able to present themselves as a supreme power, but looking back, about six decades of French rule in Indochina was at best ambiguous and, at worst, fragile.[59] This was not only because the Chinese and the Vietnamese elite continued to cling onto their precarious political hold and keep the pressure on, but also because a more entrenched cluster of powers remained in the hands of a third force. Like the Abors and Mishmis on the upper Brahmaputra along India-Tibet-China frontiers and Kachins on the upper Irrawaddy across the Burma-Yunnan frontiers, a number of local and regional power bases, including the Montagnards, "Black Flags," and "Yellow Flags," planted themselves in the transitional zones between the upstream and downstream Red River. In this context, the Red River became the site of actions by local players who Philippe Le Failler terms "traditional power and unconventional practices," and what Ella Laffey calls "local context"—key to the strengths of the riparian locals against French imperial advances.[60] This is a story worth telling.

III

The lack of a robust imperial presence in regulating the movement of people and goods within the Red River basin was informed as much by its geomorphology as by local political forces. Some of these dynamics evolved in the aftermath of the Taiping rebellion. Following its suppression by the Qing army in the mid-1860s, the rebels retreated from Guangxi, Guangdong, and Nanjing to highland Vietnam. Here they splintered into competing factions, notably the "Black Flags" army (Hei-ch'I Chün) of Liu Yung-fu and "Yellow Flags" army (Huang-ch'I Chün) of Huang Ch'ung-ying, both being the former followers of Taiping leader Wu Ya-chung. Amid this volatile factionalism, Vietnamese Kinh officials aligned themselves with the "Black Flags," while the French forged alliances with the "Yellow Flags," intensifying their confrontation.

At the heart of the factional politics and chaos in northern Indochina lay the question of control over the Red River system and its trade routes. At the beginning of 1870, the Black Flags took control of Lao Cai after displacing the Yellow Flags. By establishing themselves around the navigable head of the Red River, the Black Flags secured the crucial juncture of the flows of trade between Yunnan and Tonkin.[61] They were able to hold for several years in a historic power vacuum, when neither Qing rulers nor the Yunnan Muslims could assert authority, and neither the French nor the Vietnamese had yet established themselves. For a considerable period, the Black Flags controlled parts of the Mekong between Luang Prabang in the west and sections of the lower Red Delta in the east, as well as the Songi Canal (Canal des Rapids), the shortest route between Haiphong and Hanoi.[62]

The Black Flags' presence in both the upper and lower Red regions allowed them to dominate the highlands and monitor the entire river system down to the coastline. It took the French about a decade, a 2.5-million franc flotilla, and seven gunboats to reclaim and secure Haiphong, Hanoi, and Lao Cai from the Black Flags. Yet many parts of the lower delta remained under Black Flag control as late as the autumn of 1883. Flooding on the river, illness due to exposure to a new climate, inadequate hospital facilities, and insufficient food supply demoralized the French troops, under which circumstances "the subjugation even of only the Delta is impossible."[63]

Following the 1885 French victory over the Chinese, who supported the Black Flags, the latter's heyday ended, but the political vacuum in the transitional zone was soon filled by another set of local stakeholders. Despite the apparent stability of relations among the major powers, the French found that without the support of the local "big bosses" (*grands caïds*) of the Black River basin, it was difficult to maintain the balance of power. This realization led them to form an uncomfortable alliance with the Deo clan.

After the disbanding of the Black Flags, the Deo clan became prominent in the areas between Yunnan and Lao border. They had strategically supported the Black Flags against the Yellow Flags and, following the decline of the Black Flags, promptly filled in the vacuum, renewing resistance against the French through guerilla warfare and seeking to retain "all manner of privileges, and in particular, boat transportation rights along the Black River corridor."[64] The French decided to withdraw their army and agreed to retain such privileges for Deo Van Tri, who was in a position to sustain a "real flow of long-distance trade." The Deo clan continued to enjoy their local authority until 1930, when the last important family member passed away, and continued to assert their influence to a lesser extent until World War II.[65]

Beyond the Deo clan, many other ethnic groups continued to assert their own network along the Red-Black system, extending their influence to the Mekong Valley. Some of the groups were branded as bandits in areas such as the Red-Black divide and were blamed for disrupting the trade routes. In reality, they were likely attempting to preserve their trade networks and, contrary to the allegation of piracy and banditry, "they were only natives in revolt against the taxes."[66]

On the Chinese and Laotian borders, the Black River basin was a major site of commercial flows and resultant "local political power practices of the Tai and other ethnic groups that lasted until Vietnam took up assimilation," as Le Failler noted.[67] In February 1890, for example, a large fleet of laden junks was sabotaged, leading to the suspension of trade for three weeks, but no further such incidents were reported afterward. Officials at the Shanghai customs office suspected that this immunity from attacks was due to some "secret understanding with the pirates on that stream [the Red River] who seem to be able to retain their power for plunder, notwithstanding the repeated successful attempts of the French forces to drive them from their strongholds."[68] Sarah Turner and Jean Michaud have recently noted how, despite controlling several northern Vietnam upland trade routes, clans and armed factions like the Black Flags and Hmong sought to carve out a locally collaborative space of interactions along the riverine trade routes with those ethnic minorities not sympathetic to the Han Chinese, Thai, and Kinh trade networks. It is no wonder the commercial nodal points on the Red River, such as Lao Cai, remained the bone of contention.[69]

The layered political and commercial interests and their uneasy but mutual existence indicated the impossibility of exclusive political authority around the Red-Black systems. Despite the conflicts between the empire and the local groups, trade and commerce on the Red continued to operate through multiple stakeholders. In all these, resistance from both the weak and the strong, as well as alliance formation, were never a reflection of the geomorphological divide between the lowlands and the highlands.

The upper ranges of the rivers were the nerve centers of connections between further inland and the ocean, and these groups capitalized on the opportunities at the "borderlands before borderline," as Davis noted.[70] This heyday of local groups was the reflection more of an assertion, keen desire, and action to connect beyond terrestrial divides than the quest for self-isolation from the rest of the world. This drive and these practices were what lay at the heart of commoning at the Red riverscapes—glimpses of which are explored in the following section.

IV

As the imperial and local powers entangled, the Red-Black trade routes accommodated wide-ranging flows of everyday people and commodities, similar to other BISMRY river networks. Wars, natural disasters, and diseases occasionally disrupted the pace and extent of these flows, but rarely displaced them entirely. The interaction between humans and the physical environment was shaped by the river networks' distinct geomorphology. This landscape exposed the vulnerability of external powers while emphasizing the local community's desire for engagement with the outside world, fostering both creative tensions and collaborations.

The limitations of modern transport like steamers against rapids, bars, and other navigational difficulties meant that the boatmen of the traditional river transport system largely dominated the riverine lifeworld. These boatmen demonstrated how freedom and mobility flourished at the precarious edge of nature. However, it was not all about battling the Red's rapids and navigating the lower delta's puzzling fluidities. It was also about conversations with life, love, and longing that often transcended the immediacy of political conflicts, trade, and mundane struggles.

D'Orléans's journey on the Red River in a boat offers a glimpse into this world of boatmen. He described the boatmen as "hybrid Chinese" and recounted their tales of an island in the middle of a lake that no one could cross. The waters were "so light that a feather cast on the surface would not float," which is why they could never land on this "woman's realm," a wonderland inhabited only by women.[71] This legend reflected the boatman's separation from their homes and women, necessitating the invention of a story emblematic of a vibrant waterworld that the Red River beaconed—a water-borne materiality that transcends time and space, as Veronica Strang has powerfully described.[72] (See, for example, Figures 5.3 and 5.4.)

At Manhao, where the Red River's navigational trail ended, began another world of mundane mobility. The boat traffic was replaced by porters and mule caravans that sought the next navigable waterfront, paving the way for sprawling adventures in different inland directions. The first nodal point close to Manhao was Mengzi, whose markets were visited by "crowds of country folks" from the hill areas and by Yunnanese, many of whom had settled there and cultivated a cosmopolitan outlook near the navigable heads of the Red River.

D'Orléans met General Ma, a Yunnanese Muslim whose family hosted a feast for him and showed an "exceptionally frank manner towards the foreigners." He was friendly toward the Christian missionaries who were supported rather than harassed, although they were never converted. "You have a God," he told the priests, "so have we: we have both a book; let us be friends."[73] This aura of tolerance at the first/last navigable point of the Red River was also evident on a river bridge connecting Mengzi and Manhao. In the middle of the bridge stood a tiny temple with an "austere but beautiful" inscription and a small figure of the Goddess of Mercy (Guanyin).[74] This sense of harmony extended to sites along the river where the Chinese protected the herons, who were believed to have the power to carry the souls of the dead to heaven.[75]

In the last leg of his travel from Simao to Sadiya via the Red, d'Orléans was surprised to see the similarity in ethnicity, dress, writing, and religion from Canton to Assam, and from Sichuan to Malacca across Laos. Rather than being disparately mobile across these vast spaces, these people appeared as the members of a multiethnic family that bridged the navigable nodal points of the BISMRY rivers.[76] Ancient Khasis pioneered rice farming in the Red River Delta but moved up the river to Yunnan, Myanmar, and Assam when pushed by the conquering Vietnamese.[77] On the Laos side, the French hoped to connect Laos with the Black River, but it was found navigable for only thirty miles by light-draught launches because of barriers of rocks and cataracts. When the steamers failed to reach beyond the navigable routes, canoes did ply, and the Laotian people inhabited the "whole country" between the Red and

FIGURE 5.3: Water villages on the Red River in northern Vietnam in the early 1900s. Source: "A Meandering Photographic History of the Red River and Long Bien Bridge," *Saigoneer*, December 22, 2019. https://saigoneer.com/vietnam-heritage/24297-a-meandering-photographic-history-of-the-red-river-and-long-bien-bridge.

FIGURE 5.4: Daily life by the Red River in northern Vietnam in the early 1900s. Source: "A Meandering Photographic History of the Red River and Long Bien Bridge," *Saigoneer*, 22 December 2019. https://saigoneer.com/vietnam-heritage/24297-a-meandering-photographic-history-of-the-red-river-and-long-bien-bridge.

Luang Prabang on the Mekong.[78] Carey reported that Lolos (Yi) and numerous other communities were found "almost everywhere" between the Red and the Mekong, especially in the Sipsongpanna (also known as Chiang Hung) region.[79]

On the Black River, the majestic tributary of the Red, commercial connectivity was so extensive that social formations were influenced by the relationship between upstream (*nguoc*) and downstream (*xuoi*). The interactions along the river network generated, as Christian Lentz noted, both alliances and fissures that "shifted rapidly over time and stretched unevenly across space."[80] On the landed routes west and north of the Black River where the pull toward other BISMRY rivers was felt, another layer of interbasin mobility emerged. Taking the routes from Mengzi to Sadiya on the Brahmaputra, d'Orléans discovered a mobile world of many ethnic groups. As they arrived on the northern edge of the Black River, he encountered Laotian, Hou-ni, Lolo, and Pais people. Due to the prospect of riverine mobility, these plural ethnoscapes remained open to the world beyond their immediate physical environment. Here, d'Orléans's tour assistant, Sao, thought he could get back to Tonkin in a week on a raft, and Nam from Saigon thought he was near home, perhaps mistaking the river for the Mekong. D'Orléans himself headed to India via the Brahmaputra.

At this edge of the Black River region, between the Mekong and Red, multiple spatial imagination took place where an Annamite, a Saigonese and a Frenchman envisioned different directions and destinations, yet all imaginatively inhabited the river spaces as roads to home. For d'Orléans, the multiple possibilities of mobility were even tantamount to an exercise in the annihilation of self-identity. On leaving the edge of the Black River, he threw "a regretful glance behind," and, as India became "intelligible" to his perception, he, for that moment, "yielded to the witchery of Nirvana."[81]

The British and Chinese anxiety about the French expansion along the Red River valley, the French attempts to obtain market access to Yunnan and Sichuan, and the Chinese efforts to continue using the Red for accessing the South China Sea put the river at the center of a tripartite encounter. Despite these high-level political contestations, the Red River remained a relatively open space for the mobility of a diverse set of people and products, including the ethnic groups inhabiting the areas between the porous borders of Indochina and Yunnan. It is no wonder that, to date, there has been more trade between China and Vietnam on the Red than the domestic trade within Vietnam itself. The idea of connecting Tonkin on the Red with Assam on the Brahmaputra in the late nineteenth century made complete sense in light of this flux of people and commodities. The transregional commons, which the Red embodied within the BISMRY network, found a larger and bolder trajectory in the Yangzi. The next chapter invites exploration into the world of the Yangzi.

SIX

THE LONG RIVER
Yearning for the Yangzi

IF ALL OTHER BISMRY rivers were like high streets, the Yangzi would be a grand boulevard.[1] Descending from an altitude of about 18,000 feet, Jinsha-Jiang (Golden Sand River) flowed past Batang toward Dali and then took a sharp U-turn in Sichuan before taking the name of Yangzi, much like the Tsangpo, which became the Brahmaputra at the great bend. From the source to the sink around Shanghai, the Yangzi was powered by at least fifty large tributaries and hundreds of subtributaries joining both banks. The watery tributes from all over climaxed in summer when, as of the early twentieth century, the river rose to 90 feet in some places, drained a staggering 650,000 square miles, and poured into the East China Sea at the rate of 770,000 cubic feet per second.[2] The Yangzi was magnificent by all measures.

Yet throughout most of the history of Imperial China, the Yangzi was considered a backburner rather than its nerve center. Its commercial commodities and agricultural products, partly aided by the 6,428,000,000 cubic feet of soil-nourishing sediment annually, fed the heartland of the Chinese Empire in the Yellow Basin. With the arrival of Western imperial forces did the role of the Yangzi largely reverse, using it as a channel for drawing Chinese trade and commerce to the wider world. This shift invited renewed interest from the Chinese themselves as well as external imperial powers to scale up its transregional links as far as India and mainland Southeast Asia. Regional economic activities remained important, but the Yangzi surpassed them all.

The transregional reaches of the Yangzi were attained not merely through imperial forces and their gunboats and steamers. Local and regional political actors asserted themselves in the river network at the new wave of opportunities. Between these forces anchored along the Yangzi there were also those ordinary folks, from junkmen to petty peddlers, who made up the fluvial mundane and were an integral part of the flows that continued in new shapes and turns in the era of the Western advances. What resulted in the course of the nineteenth and first half of the twentieth century was a Yangzi that appeared as no one's exclusive zone, but a moving site of entanglements of multiple agencies whose crescendos revibrated from Shanghai to Mumbai via Sadiya, Bhamo, and Lhasa, among other nodes.

I

The Yangzi, the world's third longest river, had more connective nodal zones than other Tibetan rivers (Figure 6.1). Of the four major distinct geospatial segments, the six-hundred-mile stretch from Shanghai to Wuhan allowed the smooth sailing of blue water ships. The segment from Wuhan to Yichang, passing through a flat plain, provided easy passage for smaller draught ships. From Yichang to Chongqing, and further upstream in Yibin, where the Jinsha met the Min River (Jialing) to form the main body of the Yangzi in southern Sichuan, the river became more undulating and susceptible to onerous

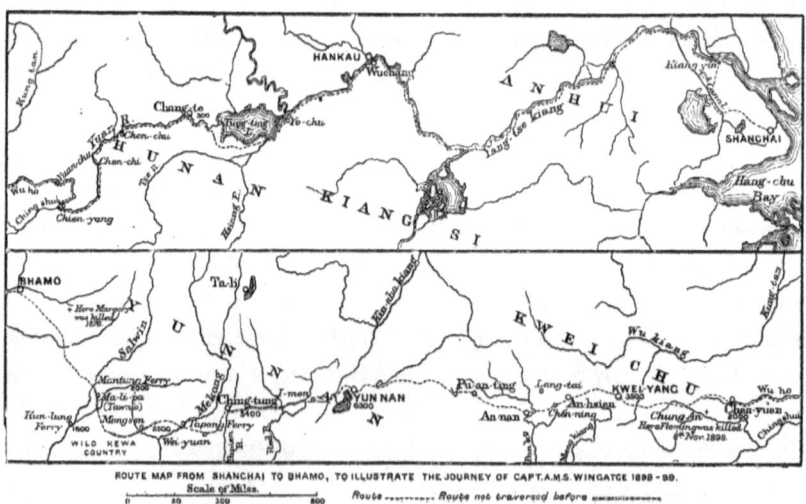

FIGURE 6.1: Route map: Shanghai to Bhamo, 1898–1899. Source: A. M. S. Wingate, "Recent Journey from Shanghai to Bhamo through Hunan," *The Geographical Journal* 14, no. 6 (December 1899), 443.

navigation due to a series of rough rapids. In its uppermost segment, from Yibin to its place of origin, the river was mostly torrential with no or negligible navigability.[3] Each of these segments provided opportunities for mobilities, but the tracts between Yichang and Chongqing were most challenging physiographically yet critically important from the perspective of transregional mobilities. This tract, stretching to Chongqing and further inland, therefore, represented the depth and breadth of human labor invested in overcoming extreme geographical barriers to reach the upper Yangzi and, thereby, Sichuan, Yunnan, mainland Southeast Asia, northeastern India, and Tibet.

The extraordinary transregional push along the upper Yangzi was deeply linked to the political dynamics in lower Yangzi, especially Shanghai. Shanghai, like other Asian ports such as Chittagong, Rangoon, and Saigon, experienced a revival in the context of Western encounters. Its rise coincided with the Yangzi's transregional thrust, partly driven by the discontent over the Yellow River, which was "rapid, tortuous, and turbid with mud." The Yellow River's elevated bed above the flood level made navigation arduous, and the constant need to manage its shifting banks drained the imperial treasury for centuries.[4] These challenges were further compounded by the uncertainties regarding its connection to the lower Yangzi via the Grand Canal. As a British merchant noted, when the Western powers blockaded the mouths of the Grand Canal, it was like putting their "hand upon the throat of the [Qing] empire."[5] By the early nineteenth century, therefore, China desperately needed the middle and upper Yangzi Valley for economic survival and political breathing space.

Beyond Shanghai, the Yangzi's transregional outreach unfolded in some of the nodal zones, with Wuhan being a major one. Wuhan, located at the Han River's confluence with Yangzi, formed a "vast interior delta"—a region that, when considered along with Hunan's Dongting Lake and Xiang River valley to the south and the Han River highlands to the northwest, constituted a "coherent and interdependent ecological system."[6] The emerging nodality of Wuhan was reflected in the fact that, in the six decades since the establishment of the Maritime Customs in the 1850s, its annual trade value rose from about 15,000,000 Haikwan taels to over 200,000,000 taels. Shipping enterprises showed equally impressive expansion, featuring 58 steamers and some 300 steamships operating on the eve of World War II.[7]

Wuhan's growth was as much linked to Shanghai as to the nodal points further upstream and a connected network of land routes. The Mandarin Road, a combination of forty days of travel by boat and sixteen days on land,

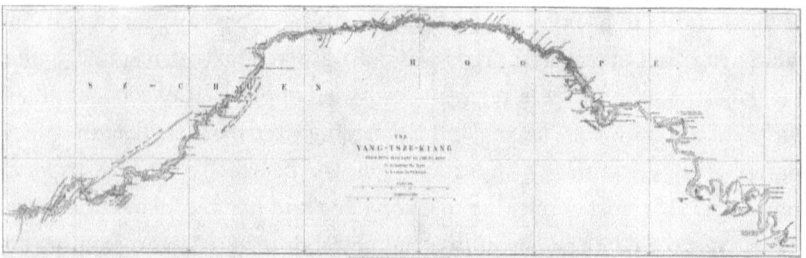

FIGURE 6.2: The Yangzi elevation between Yichang and Chongqing. Source: R. Swinhoe, "The Yang-Tsze-Kiang from Tung-Ting Lake to Chung-King to Accompany the Paper by R. Swinhoe," *Journal of the Royal Geographical Society of London*, 40 (1870).

connected Wuhan with Kunming.[8] Another route, about a thousand miles long, connected Wuhan with Bhamo on the Irrawaddy via Yochow (Yueyang), Dongting Lake, and Tengchong. A far more distant connection was with Mumbai, which was considered accessible from Shanghai via Wuhan with a combination of land and river routes and a proposed railway. The maritime distance between Mumbai and Wuhan via Shanghai was 5,400 miles, but a voyage from Mumbai to Rangoon (2,150 miles) and then overland to Wuhan via Chengdu, using the Irrawaddy-Yangzi route, would save about 1,650 miles of oceanic travel.[9]

On the main stream of the Yangzi, Yichang was the next nodal point. Dubbed the "gateway to the west," Yichang was principally a transitional zone between Wuhan and Chongqing, marking the end of soft-sailing and the beginning of a montane landscape with tougher navigation[10] (Figures 6.2 and 6.6). It offered a glimpse of the changing geomorphology that became more pronounced in Chongqing. In 1913, British botanist E. H. Wilson noted that the Min river and its network, including its three major navigable tributaries (Fu, Ba, and Zhou), drained half of the Sichuan province (more than 40,000 square miles) into the main body of the Yangzi. Even before joining the Yangzi, the Min, Wilson observed, was the "greatest collecting and distributing waterway in Szechuan, and its commercial importance was probably greater than that of the Yangtsze itself." Similarly, the Yalong River was noted by Wilson as an "artery almost equal in volume" to the Yangzi. Other rivers on its right bank, joining from Yunnan and Guizhou, were equally significant distributors of commodities.[11] This "confluential" nodality, as a "collector and distributor to Western China," led to comparisons of Chongqing with London in relation to Europe prior to the opening of the Suez Canal.[12] By the turn of the twentieth century, over 10,000 junks passed annually between Chongqing and Yichang, each carrying 50 to 100 tons of cargo (Figure 6.7). As Archibald Blue noted, this was about twice the freight for the 11,000 miles between Liverpool and Shanghai.[13]

Of the total recorded export from Chongqing, local opium comprised about 45 percent, white wax 15 percent, silk products 12 percent, medicine 10 percent, and musk 8.5 percent. The rest comprised bristles, hemp, mushroom, brown sugar, feathers, leather, safflower, turmeric, chitan root, paper fans, and cuttle fish. Official statistics only captured a fraction of Sichuan's export, especially that of silk whose production was "practically unlimited." Lead and copper from the Yunnan mines were also part of this export range. The import items were dominated by Indian cotton yarn (46 percent), followed by shirting, including those of Italian origin (26.5 percent), American clarified ginseng, aniline dye, seaweed, and agar-agar, among other items.[14] The annual value of the total recorded trade of Chongqing averaged around £5 million sterling between the late 1870s and the eve of World War I.[15]

The extent of the cotton trade was reflected in the scraps of cotton on the roads and stones from the Irrawaddy to Yangzi Valleys and in the strings of porters struggling along under the huge, unpressed bales, looking like "ants under their eggs in the breeding season."[16] An extension of trade flows on the Yangzi above Chongqing was visible in Yibin, the last major navigable point on the river. Located at the confluence of the Min and Yangzi Rivers, the greater Yibin area produced raw silk and various silk products, coal, and salt. Import items included European goods, cotton, cutlery, toys, and "every bad bauble that Leipzig or Nuremberg can export."[17]

The commercial vibrance that reached Chongqing, despite extreme navigational challenges, was prominently felt across Yunnan and Sichuan. A considerable amount of locally produced products in Sichuan and half of all the products imported to Chongqing from the lower Yangzi went to Yunnan.[18] Nanny Kim offers an exhaustive account of commercial routes connecting the upper Yangzi Valley across Sichuan and Yunnan and provides clues to further transregional connections beyond Yunnan's frontiers.[19]

Among many such routes, the northernmost was between Chongqing and Batang in Tibet (Figure 1.4). Batang, a fertile plainland at the crossroads of Chongqing, Chengdu, Lijiang, Dali, Sadiya, and Lhasa, was situated on the historic Tea Horse Road between Tibet and Chengdu, overlooking the Jinsha.

It was also a starting point of the 180-mile road to Sadiya on the Brahmaputra via the Zayu River basin spanning Tibetan villages, monasteries, and partially the basins of the Salween and the Mekong. The route took about twenty days by pack mules and was open except for about four months in winter.[20]

Dubbed the "Eden of Eastern Tibet" because of its abundant crops, vegetables, fruits, meat, and fish supply, Batang saw "thousands of yaks and mules" annually cross the Himalayas on their way to Lhasa laden with tea.[21] Part of this traffic, especially comprising the Bhutanese and Tibetans, came down to Tezpore in Assam on ponies and mules, thus making a triangular trade network between India, China and Tibet.[22] Cooper met a Chinese merchant near Batang, hailing from Zayu "trading post," who described Batang as leading over four great rivers, viz., the Tsangpo, Salween, Mekong, and Jinsha—an observation that testified to the centrality of transregional land-river route.[23]

Another route from Yibin connected to Lijiang, about four hundred miles upstream. Part of the Silk Route that connected Chengdu, Lijiang was a center of tea and cotton trade, exchanging these goods for musk, wool, fur, gold, goat skins, and medicine. Its tea trade with Tibet was valued at a million taels.[24] Lijiang was also a key point where connections to India were made via the Brahmaputra, as noted in chapter 1.

Thomas Blakiston, an English explorer and naturalist, hoped that one day steamships braving the upper waters of the Yangzi would be met in Yibin with travelers from Calcutta who would "save the time and inconvenience of the sea-voyage by way of Singapore, coming on overland route through Burmah to meet these steamers in Yunnan."[25] This ambition was shared by Henry Cotton and T. T. Cooper, who envisioned connecting Sadiya on the Brahmaputra to Lijiang on the Yangzi. During his trip in 1868, Cooper found no greater obstacles on his first leg of the journey between Yibin and Kiating than those from Yichang to Chongqing. He noted that this route was part of the Chengdu-Assam route, the Indian portion of the Silk Route. From Kiating, the route entered India via the Brahmaputra Valley through upper Burma, completely circumventing the Tibetan plateau.[26] The vitality of the route connecting the Yangzi and the Brahmaputra was evident in the constant movement at the ferry in Taku village near Lijiang, where crossings occurred even in the freezing cold of December, often using boats crafted from inflated sheepskin.[27]

The Yangzi's outreach to the Irrawaddy was mainly through Bhamo. Chongqing was nearly equidistance on the 1,500-mile-long route between Bhamo and Shanghai, and was remarkably alive with human and commodity flows

(Figure 6.3). One example of such mobility relates to Morrison, the Australian traveler who crossed the distance in exactly one hundred days. He "purposely waited till the hundredth day" to complete the journey, which cost him only $100.[28]

Silk in "very large" quantities came from Sichuan to Yunnan, with some of it still making its way to Burma via this route.[29] In 1891, it was reported that Chongqing's imports and exports totaled 11,736 tons, most of which went to Yunnan and the rest to Burma via Bhamo.[30] By the turn of the century, Dali had become a center of calico weaving by Sichuanese weavers, using Bombay yarn that arrived from both Shanghai and Bhamo. The supply from

FIGURE 6.3: Map of a journey between Shanghai and Rangoon. Source: George Ernst Morrison, *An Australian in China: Being the Narrative of a Quiet Journey Across China to British Burma* (Horace Cox, 1895). Last page. © Staatliche Kunstsammlungen Dresden, Museum für Völkerkunde Dresden / Eickstedt, Egon von. All rights reserved. https://digital.staatsbibliothek-berlin.de/werkansicht?PPN=PPN618598049&PHYSID=PHYS_0368&DMDID=DMDLOG_0053.

the Burma side was so substantial that there were suggestions to set up a buying house for Bombay yarns at Tengchong.[31] All these made Dali a center of manufacturing, with its linkages supported by supplies from both India and China.

Between 1890 and 1898, exports and imports to and from western China more than doubled, increasing from 16,218,400 rupees to 39,579,400 rupees. If the Shan States were included, the total trade in the same period would have increased almost fourfold, from 55,426,300 rupees to 193,587,300 rupees. The products included raw silk, hides, opium, orpiment, horns, fibrous products, and miscellaneous items such as gold and silver, brass gongs and pots, iron cauldrons, straw hats, paper, hams, musk, fur coats, walnuts, china root, coptis root, copper, white wax, wool, and medicine.[32]

Most of the products that moved from the westernmost navigable part of the Yangzi to the Irrawaddy transited through Yunnan, using five main routes: Wuhan to Kunming; Chongqing to Kunming via Kueiyang-Fu; Chongqing to Kunming via Luchow; Yibin to Kunming; and Chongqing to Dali via Chengdu. Eventually, most of the products found their way to different parts of the Irrawaddy, predominantly Bhamo.[33] Except for the years of the Hui rebellion, the annual trade traffic between Dali and Bhamo was consistently high, with the value of the trade estimated at 500,000 taels in the 1890s.[34] This brisk transregional trade between Burma and western China declined during the Chinese Revolution in 1911 but recovered in the following year, with an increase of 40 percent to 640,000 rupees.[35]

The social and economic changes during late Qing period, along with transport developments in the mid-nineteenth century and significant population growth, led to the enhanced use of the overland and river routes involving various agencies, systems, technologies, and organizations. A certain degree of professionalism and kinetic forces of humans and animals worked together to further dismantle the geomorphological divide along the navigable and nonnavigable Yangzi, cementing a network of transregional connections. Mules and horses were shod, better services were provided along the road with food and shelter, and pack animals were trained to be more efficient members of the caravans.[36] The nodal points of these intervalley routes connected Kunming and Dali—a route that went "chiefly along the headwaters of a series of streams, which flow north and discharge themselves into the Yangtze."[37] Efficiency on the land route meant that the trade potentials of the rivers were realized to a greater degree.

The human and commodity flows between Shanghai and the routes that connected Indochina, Burma, India, and Tibet were so much entrenched in the Yangzi network that Morrison suggested that "no man can venture to assert that any other trade route (across Sichuan) exists, or can ever be made to exist, than the River Yangtse; and all the French Commissioners in the world can no more alter the natural course of this trade than they can change the channel of the Yangtse itself."[38] The centrality of the Yangzi, however, should not obscure the fact that the river was a collection of tributaries connecting it to the interior. Nearly every town and notable village stood upon these confluences, where "Here and there men will be found hauling small, stout-bottomed boats over the stones at the mouths of these small rivers."[39]

In this broader context, one can appreciate George William Skinner's suggestions that China was neither a single national economic system nor a set of provincial economies, but rather a collection of eight "macroregions" of trade, commerce, and mobility. Each macroregion was marked by a range of centripetal economic activities that followed a water transport network spanning a core and its tributary peripheries.[40] These macroregions were separated by the high cost of transportation before the arrival of modern communications.

While Skinner's framing of the macroregion is generally accepted by economic historians of China, the vantage point of the Yangzi questions this theory. Out of these macroregions, three were directly connected to the Yangzi, including Chongqing. The gravity of the Yangzi as a networked transregional site defied any spatially autonomous economic organization. Chongqing became a major urban center even though reaching here from the middle Yangzi was extremely challenging until at least the late colonial period, even when modern transport like steamers began to ply the river. The transregional flows continued along the Yangzi after it ceased to be navigable by large boats. In this context, Chongqing served as a signpost and springboard for mobility across Sichuan, Yunnan, and beyond on the one hand, and Wuhan, Shanghai, and the East China Sea on the other. External imperial activities further fueled these transregional mobilities and flows, although it was the Yangzi rather than the imperial forces that continued to dominate the life and landscape of this transregion. The human physical and emotional energy, financial investment, sheer optimism, and imagination that poured into the efforts to connect the lower Yangzi with Chongqing and beyond via a precarious watery route, as detailed in this chapter, challenge the idea of spatially bounded centripetal organization of economic activities.[41] This is precisely why transimperial rivalry unfolded along the Yangzi's diverse geomorphological features, shaping an expansive spatial itinerary.

II

From the mid-nineteenth century until World War II, commercial nodal points along the Yangzi network became the sites of intense entanglements among several empires. The British announced their presence in the lower Yangzi with unmissable alacrity following the Opium War, but other imperial forces quickly followed, prompting the British to seek fresh pathways along the upper Yangzi, Irrawaddy, and Brahmaputra. By the late 1860s, Edward Sladen, a British political agent at the court of the last Burmese king in Mandalay, was worried about the American advance on the east coast of China and expected further inroads following the construction of the Pacific railroads. Because of this "contingency of US predominance" impacting the British trade in Shanghai, and in the context of the decline of the opium trade along with the Canton system, Sladen pointed to the need to find a western doorway to China.[42] From the British perspective, this was a reasonable call because, by this time, the Brahmaputra and a third of the Irrawaddy Valley were formally under the British sway.

The British attempts to connect the Yangzi to the Irrawaddy gained momentum with the exploration led by Augustus Margary, a young British diplomat. In 1875, Margary led his team from Wuhan toward Bhamo, while another team led by Browne started from Bhamo toward Tengchong, the designated meeting point between the two parties.[43] However, whereas Cooper's attempt to link China and India via the Yangzi and Brahmaputra was a frustration, as discussed in chapter 1, Margary's attempt to establish an imperial connection between the Indian Ocean and the Pacific via the Yangzi-Irrawaddy network ended in tragedy. He was murdered in the Tengchong area by some irregular Chinese troops who believed that Margary was planning a railway network.

The British used Margary's death as an excuse to establish a greater English presence in the upper Yangzi, leading to the Shefoo convention in 1876, which compelled China to concede to the British the right to trade in Chongqing.[44] With the annexation of the remaining northern part of Burma, including Bhamo, in 1885, it seemed that the British were poised to strengthen their imperial connections between the Yangzi and the Irrawaddy Valleys.[45] They also managed to minimize the Russian presence in the Yangzi Valley by agreeing not to seek railways north of the Great Wall in exchange for Russian abstention in the Yangzi Valley.[46]

Despite these geopolitical gains, the British-sponsored transregional connections never fully materialized along this route. The lack of profitability due to high transit costs, excessive insurance premiums, and *likin* charges forced British traders to conduct business through Chinese merchants in Sichuan. In fact, as Archibald Blue noted, the Americans profited more than the British as they sailed steamships on the Yangzi using their navigational skills gained from the Hudson and Mississippi Rivers. American steamers, for example, could make two trips per month between Shanghai and Wuhan, while the British would take more than a month to do so.[47]

The French influence in the middle Yangzi never surpassed that of the British. By the turn of the century, the French contemplated constructing a railway from Yichang to Chongqing and purchased a few miles of the Yangzi foreshore at Yibin through the help of Jesuits.[48] This idea demonstrated the French appreciation of the commercial importance of the Yangzi network in Sichuan, but beyond their contested presence around the Mekong and Red River Valleys, they achieved little tangible success.

The German presence on the Yangzi was similarly uneven. They joined forces with other imperialists, using gunboats (*Flusskanonenboote*) to secure Hubei, Sichuan, and Yunnan from the so-called river pirates (*Flusspiraten*).[49] The German presence in the upper Yangzi was also facilitated by the British merchants' relative lack of enthusiasm for outreaching to those regions. Additionally, German banks' advance financing enabled them to penetrate the Chinese interior, and they generally ended up representing British firms.[50] However, the apparent success of the Germans was punctuated by losses and accidents. In December 1900, the German steamer *Suischang* wrecked on her maiden upper Yangzi voyage. It collided with a submerged rock at the Kung Ling Rapid, thirty miles above Yichang, resulting in the total loss of the ship as well as the deaths of a few Chinese crew members and the captain. The remaining German interest never caught up with that of the other contenders.

The disjointed ventures of the European powers in the Yangzi Valley, each seeking to outmaneuver the other, were overshadowed by the Japanese advance that became evident after their startling win over the Russians in 1905. Britain was willing to admit new Japanese enterprises in the Yangzi in exchange for the Japanese allowance of British operations in Manchuria.[51] By the start of World War I, Japanese products were already a "serious rival" to

European and American products in the Chinese market. The British (including British India, Hong Kong, and the Straits Settlements) still commanded about 50 percent of Chinese international trade, but Japan and Korea were fast catching up.

A British merchants' association claimed that the Japanese influence in the Yangzi Valley rose due to the investments that initially originated from Britain. On the eve of World War I, the British sought "Japanese recognition of their claim to the privileged position in the Yangtse Valley," having "nothing to bargain with [Japan] except the official offer of financial assistance in the future."[52] In the decade leading up to the war, the tonnage of Japanese shipping in Shanghai rose from 14 percent to 22 percent, while British tonnage fell from 48 percent to 40 percent. At Wuhan, British tonnage rose by 6 percent, while Japanese tonnage rose by 13 percent. Overall, the total tonnage of Japanese steamship companies equaled that of the British, at about 250,000 tons—making it about three-fifths of the British tonnage in Shanghai and more than three-fifths in Wuhan, with a similar trend in the upper Yangzi.[53]

British acquiescence to Japan's commercial rise was largely tied to its military ascendancy. In Wuhan, where they already had a military base, the Japanese showed an active interest in constructing Nanchang-Hankou line. The Qing government could hardly obstruct any Japanese enterprise in those parts of the Yangzi basin where British presence was not established.[54] The events leading up to World War II shifted influence in southeastern China from the British to the Japanese, in the Yangzi Valley where Japan "had her stiffest fence, and the biggest prize." The Japanese successfully and quickly denied third powers "their indisputable right of navigating the River" more rapidly than in Manchuria.[55]

In response to these external pressures and its waning clout in the lower Yangzi, the Qing government opted for what was its only choice: to keep a check on the external influences on the upper Yangzi, making Chongqing the next Chinese bastion to protect its interest. In this context, China was hardly ever "a sleeping lion," especially considering its mobilities through western China, mainland Southeast Asia, India, and Tibet. Rather than a sleeping lion, China was a prowling cat, with the Yangzi offering a crucial transregional window toward its western frontiers. Despite varied geomorphology and navigational difficulties in its upper reaches from Yichang onward, this mobility reverberated beyond its last navigable points, continuing to foster connectivity, both real and imagined, across a vast territory of Asia.

In this general scheme, the stretch of the Yangzi from Yichang to Chongqing was particularly sensitive for both the late Qing and Republican China. The right to navigate the upper Yangzi, especially in Chongqing, was "only conceded by the Chinese Government after great diplomatic pressure," wrote a secretary to the London Chamber of Commerce to the British foreign minister.[56]

The key to China's survival strategy was to delay steam shipping on the upper Yangzi. One tactic was to allow the British merchants to ship their products between Chongqing and Yichang only on hired Chinese-built boats or boats built in Chinese style. Archibald Little's *Kuling*, designed to run through the rapids along the gorges, was not allowed entry into Chongqing. In fact, the Chinese forced him to sell it to them in 1888 for three times the construction cost of the steamer. While the Chinese used the *Kuling* between Wuhan and Yichang, they made Little agree to stop navigating the upper Yangzi for the next ten years.[57] Little, who believed there were no gospels outside of Christianity and free trade, was determined to overcome Chinese resistance. He eventually concluded that, in terms of granting access to steamers, Chinese officials displayed "disheartening effects of perpetual procrastination, an art in which the Chinese are admittedly 'Past Masters.'"[58] While the Chinese impeded faster British imperial mobility along the upper Yangzi, they advanced with their dual program of westward expansion, which included civilian resettlements from Sichuan to Rima region. This progress alarmed the British to such an extent that they feared that the Chinese would take over the tea gardens north of the Brahmaputra Valley.[59]

When a complementary agreement to the Shefoo convention in 1891 proclaimed Chongqing as an open treaty port, the operations of the British steamers were made contingent upon the launching of Chinese steamers first. By preemptively retaining the first right to running steamships in Chongqing, China secured "very much the best of the bargain."[60] In delaying the steamer operations on the upper Yangzi, Chinese Yamen officials argued that steamers shouldn't run on the Yangzi between Yichang and Chongqing for two grounds: first, it would lead to clashes with local junks and their loss; second, many local junk crews would be left unemployed, leading to widespread discontent. These assumptions were emboldened by the prediction that local crews and trackers would deliberately sink the steamers and that monkeys would hurl rocks on the decks, which the Chinese could not prevent.[61] They also took recourse to a story in which Emperor Yu of the second century BCE, while opening the channels through the gorges, deliberately left the rocks to prevent ships from running through this part of the upper Yangzi. Such Chinese resistance to external influences in upper Yangzi ensured that no regular steamer from Yichang to Chongqing would run until as late as the 1920s.

From the 1850s through the 1940s, the Yangzi thus saw the rise and fall of various powers that were temporally unstable and struggled to maintain control over different fluvial nodal zones of the river. The ensuing contestations, contacts, and compromises led to the emergence of the Yangzi as a "legally" shared space shaped by diplomatic and military pressures but truly guided by the logic of the river's trade and commerce flows. By 1910, there were forty-six open ports in China, most of which were on the Yangzi. These ports were "chosen for the role they could play in extending or regularizing the shipping network rather than for their own potential as trade centres," as Anne Reinhardt noted.[62]

Beyond the stories of the rise, fall, and unstable assemblages of empires along the Yangzi, there was another layer of contestation between the Chinese empire and the locals who called the Yangzi Valley their home. This story of encounters centered on the Yangzi, ranging from everyday passive resistance to armed rebellion, is important to understand the river's multifaceted agency. This agency allowed it to accommodate unconventional political forces, as was the case with other BISMRY rivers.

III

Between 1850 and 1864, the Taiping Rebellion shook the Qing Empire in a considerable show of strength. The Taiping rebels' entanglement with the Yangzi must be understood in the context of their quest to carve out a space within the new economic opportunities as discussed earlier. Like the Faraizis in the Brahmaputra-Ganga Delta and the Black Flags on the Red River, the Taipings staked their fortunes at the commercial crossroads of the Yangzi.[63] Whereas the Faraizis promoted a rational-puritan Islam, the Taipings anchored their movement on a populist version of Christianity.

Their movement began in Guanxi and quickly spread downstream along the Yangzi to Nanking, where they established their headquarters in March 1853. The rebels and imperial forces fought in cities and towns along the river banks as well as at the Grand Canal's confluence with the Yangzi.[64] Hong Rengan, the Taiping leader, and his supporters' original strategy was to "set a foundation in Nanjing, expand their reach down the Yangtze River to

Zhenjiang to seize control of the Grand Canal and claim Anqing upstream to control the upper reaches of the Yangtze," as Platt describes in an interesting study. Following this, he suggests, they would "consolidate the seven southern provinces, campaign westward to take Sichuan and Shaanxi, and thus would the kingdom be established—a southern empire stretching from the Yangtze River to the ocean."[65] Their fight for nodal locations and cities along the Yangzi aiming at securing the tea- and silk-producing areas and controlling the fertile lower reaches of the river known as Jiangnan (literally, South of the River) was consistent with their strategic goals. The perception that the Taiping capital of Nanking was "the city of the Coolie Kings" and that their chief customs official was originally a "common coolie" also indicated a Taiping polity that resonated with the Yangzi as a site of economic activities.

Behind this strategic move around the Yangzi, however, lies a deep connection to the lifeworld that the river provided. This is reflected in the conceptualization of the river as a "living creature, serpent, with his head at Hubei, its body in Anhui and its tail in Jiangnan" around Shanghai.[66] True, the Taipings were not originally rooted in the middle Yangzi basin, only expanding their influence from Guangxi. But this mobilization appears to be a translocal political project that followed the rise of the Yangzi as an emerging economic belt, gradually overshadowing the post–Opium War Pearl–West River network. While the Yangzi—as a corridor of mobility, flows, and frictions—invited even such groups as Taipings, who were not local per se, their remarkable political and cultural alignment with the river against powerful opponents also speaks to a process of localization.

The Taiping rebellion was a quest to establish dominance over the Yangzi as a site of renewed commercial buoyancy, a goal shared by both imperial China and European forces. Consequently, the Taipings' fate was ultimately decided by a transimperial coalition against them. The Qings sought to clear the Taiping from the Yangzi as early as possible and succeeded with the support of European forces who, despite their Christian persuasions, shared the Chinese imperial goals toward a "peaceful" Yangzi.

The defeat of the Taiping brought an end to the era of Yangzi-based regional political formation and paved the way for clearer encounters between external imperial forces and the Qing. But the Taiping was one of several manifestations highlighting entrenchment on the Yangzi as a commercial network. If the Taiping rebellion underscored the localized stakes in the middle Yangzi, the Hui rebellion, which evolved around the same time and outlasted the Taiping, reflected their commercial interests in the upper Yangzi and other parts of the broader BISMRY networks. David Atwill details the political and military conditions of the Hui in western Yunnan.[67] A similar exploration in eastern Yunnan and Sichuan would reveal a considerable Hui presence along the upper Yangzi nodal points. Although most parts of the Jinsha Jiang were not navigable, its banks were busy trade routes, and Hui settlements were common along the tributaries of the Yangzi all the way to Ichang. To hold sway in imperial China's southwestern region, the Hui needed to secure the upper Yangzi trade routes.

Blakiston's note on the presence of a thousand Muslims in Chengdu was probably an understatement. Curiously, the entrance of the first gorge on the Yangzi above Yichang was locally known as "Mussulman Point."[68] The term was also mentioned in a writing by another British traveler, Percival, who described arriving at the "Mussulman Point" where the river "suddenly changed its course to the west. As soon as the corner was passed, the river narrowed from over half a mile to about two hundred yards, and the gorge instantly burst upon us in all its wild and savage grandeur. What a grand, what a glorious site!"[69]

It is not clear whether the "Mussulman Point" indicated a significant Muslim political or economic presence in the upper Yangzi, but it is plausible that the Hui attacks on Chengdu and Pingshan were encouraged by the considerable existing clout of the Muslim population.[70] The influence of the Muslims in Chengdu was reflected in the fact that T. T. Cooper, on his journey to India in early 1868 via the Yangzi and Brahmaputra Rivers, had a stopover for "consultation" with a Muslim Mandarin.[71] Overall, one account from the 1920s indicates that Sichuan had more Muslims than Yunnan, with about 700 mosques, 70,000 families, and 400,000 individuals. As Davenport observed, if there had been no internal conflicts among the Muslims of the region, the Chinese would not have defeated them and prevented them from setting up a "great Mohammedan State in Western China."[72]

It took the combined forces of the European powers and Qing China to crush both the Hui and the Taiping. The Yangzi was sliced into various legal entities to accommodate different imperial powers, but the political ambitions of both the Hui and the Taiping lingered for decades to come. Such ambition persisted as late as the 1890s when, in the name of resisting the Japanese, the Hui armed themselves to revive their political hope.[73] At the turn of the century, half the population in Yunnan and Sichuan were non-Chinese, and along with other ethnic groups, Muslims were still an "extremely strong force."[74]

In addition to the two most notable Yangzi-based manifestations of political assertion, smaller resistance around the strategic and commercially important nodal points on the river continued. Pingshan City, for example, "often serves as a battleground between the Chinese and the Lolo tribe," who inhabit a vast expanse of land to the west along the Yangzi tributary network, where the rugged terrain has enabled them to maintain their independence.[75] George Litton identified the Lashi people as closely related to the Mossos of the upper Mekong, originating in the greater Lijiang region, and noted that they "entirely" monopolized the Yangzi bend.[76] The Lashis, who resembled the Abors of the Brahmaputra bend, were among many "semi-Tibetan" people who were politically active against the Chinese. Some European sources noted rumors in Chengdu about the Chia Kung States which, if true, "foreshadowed the coalition of all the tribes from the Min to the Mekong in one great upheaval against Chinese suzerainty."[77]

There were eight such Chia Kung States that occupied the mountainous tract on the right bank of the Min River, with their territories stretching from Sungpan Ting in the north to Kuanhsien in the south. One prince in the region (of Wassu) refused to pay tribute to the Republican government in Beijing, claiming that China could only be governed by a strong hand.[78] Republican China's attempt to secure the valley between Yichang to Chongqing zone was only cloaking the real power-holder in that region. Between 1927 and 1934, Liu Xiang and his 21st Army controlled this Yangzi territory with "complete autonomy" from the nationalist government downstream and "effectively diminishing some of the treaty privileges" to the external forces.[79]

As late as the 1950s, American and British intelligence officials believed that the region south of the Yangzi might break away from the rule of the Chinese Communist Party in the "not-too-distant future."[80] Although this proved to be incorrect, it was based on the perceived resilience and influence of local political forces up to that time. All these ethnopolitical assertions, including the armed rebellions of the Taiping and the Hui, indicate that local political forces were looking at the world via the Yangzi. Rather than being cowed by the imperial forces of the plains, they went on the offensive, using the river as a strategic point to engage with the ocean and the world beyond. These disruptions of the lowland-highland dichotomy embodied a form of rooted mobility within the Yangzi network, fostering deeply embedded glocal flows of people and commodities that sustained a remarkable process of commoning.

IV

In the early 1860s, Blakiston envisioned asserting influence over the Yangzi and working to "humanize Celestials" through mercantile enterprise, relying on the "powerful agency of steam" rather than military force or missionary efforts.[81] Echoing Blakiston, a few decades later, a British consular agent in Chongqing asked, "If a boat drawing four feet can be dragged over the rapids by a hundred men and boys, half of whom merely shout and leave the pulling to the other half, what is there to prevent steamers of special construction, of equal draught, and with a steam power exceeding the strength of half a hundred men and boys, from ascending?"[82]

Yet neither the railway nor the steamers could match the sheer audacity of the upper Yangzi, while its economic allure was too strong to resist, ensuring relentless mobility into that region and beyond. This reality guaranteed the ongoing local participation in the Yangzi lifeworld. Steamers could comfortably come up as far as Yichang, but the onward journey to Chongqing and Yibin was an uphill task due to untamable rapids. As late as the turn of the twentieth century, there were only ten steamships on the middle Yangzi, and fewer upstream. Consequently, the bulk of transportation still relied on junks. In 1913, Chongqing saw 13 steamships with a tonnage of 2,548 compared to 776 chartered junks with a tonnage of 23,718.[83] Local junks and human energy continued to dominate the Yangzi lifeworld, where boatmen enabled, as Blakiston himself noted, "river navigation to perfection" that beat British or even North American boat voyaging "all to pieces."[84]

With "flat bottoms, square bows, and turned-up sterns," like those running on the upper Mississippi, these boats were especially suited to navigate the rough water through rapids. Of the 48 types of local boats that commuted through this challenging tract, 46 originated in Sichuan and 2 in Yunnan, and their capacity varied from 20 *piculs* to 2,000 *piculs*.[85] Steamers and plain sailing boats transported goods 1,000 miles up the river to Yichang, from where about 7,000 junks made round trips annually to Chongqing.[86] At Yichang, a British military official, Henry Sarel, saw an "immense number" of junks moored along the bank during his visit. Similarly, in Shashi, in the district of Jingzhou in Hubei on the left bank of Yangzi (between Wuhan and Yichang), Blakiston saw junks closely packed along the river bank for nearly two miles, as this was a transit point from and to Chongqing.[87]

The relative sluggishness of mechanization of the transport system meant that, like the rest of the BISMRY rivers, the Yangzi continued to shape the mobility. While political forces sought to secure the nodal points, the drivers of everyday mobility—boatmen, porters, and travelers from far and near—connected with the longer lore of the river. Bridging the different geomorphologies of the tract across Yichang, Chongqing, and Yibin was a testament to human labor at its peak and an immense depth of imagination of the fluvial world of the Yangzi.

At least 100,000 boatmen, involved in long-distance shipping through the Three Gorges between 1750 and 1850,[88] carried not only merchandise but also their skills, pains, pleasures, and aspirations tied to the flows of the Yangzi. The lifeworld created by the symbiotic relationship between the boat crew and the extremely challenging landscapes of the upper Yangzi is well worth exploring for at least two important reasons: to understand how the Yangzi encompassed power and powerlessness, commercial prospects, and physical challenges leading to a fluvial common space of interactions among imperial power, labor, and society; and to examine how human labor was practiced in this tough Yangzi corridor between Yichang and Chongqing, which provided a passage between the South and Southeast Asia and cemented a transregional space of connectivity across southern China.

In many cases, human labor was exploited to the extent of depravity. Since their emergence in the wake of the late Qing commercial boom, the Yangzi boat crew became social pariahs, often subject to oppression and ill treatment by boat owners. They were poorly clothed, poorly fed, and away from families for much of the year. Yet the work life of the Yangzi laborers was more nuanced than a simple story of labor exploitation. The people, commodities, and inspirations that moved across this segment of the river reflect the vitality of Chongqing, a nodal zone mediating mobility from the lower and middle Yangzi to the upper Yangzi and beyond. The Chinese imperial network, external imperial forces, and other regular travelers relied on these boatmen, who embodied the magnitude of the Yangzi and the mobility that it inspired.[89] Their labor represented their marginalization in society, but it also reflected the empire's dependency on them on multiple fronts. There was an element of self-exploration, empowerment, and entanglement that can only be understood through exploring the inner life of the Yangzi laborers (Figure 6.4).

The work pattern of the Yangzi boat crew, Chabrowski noted in a fascinating study, reflects a mnemonic practice that synchronized their physical

FIGURE 6.4: Trackers hauling a junk over the Yeh-Tan (Wild Rapid), upper Yangtze River (1905). Source: University of Bristol—Historical Photographs of China, reference number: CP-s011. An unposted post card. FD-s309. https://hpcbristol.net/visual/CP-s011.

attention and movement with the difficult places they passed. Their songs, *haozi*, represented their deprivation and isolation from family but also served as a real-time mnemonic technique to negotiate the tough, rapid terrains. The songs, in their bare essence, became a list of places and their characteristics. As Chabrowski notes: "Limiting the number of objects to remember and sequencing them according to performed activity allowed for a narrowing of reality and for organization of it in a relevant manner. A boatman could thus recall an appropriate station and plan his movement through the world of the rivers as particular steps on his journey came under his mental control."[90] Blakiston equates this practice of running ahead and going down on their knees, "chin-chin" them, with praying at the altar in Buddhist temples.[91] In this context, the shape of the gorges through which the rapids ran evoked different kinds of organic imageries: Yellow Cat, Ox Liver, Horse's Lungs, Buffalo Mouth, Witch's Gorge, and so on, following the formation of the rock. These were not just rocks, but markers of lived mobility.

If the gorges stood as time-stamps and markers of mobility for the Yangzi boatmen, for the Europeans on the move these were nature's wonder whose sublimity transcended immediate worldly gains and pursuit for imperial

FIGURE 6.5: Entrance of the Lu-Kan Gorge, upper Yangzi. Source: Thomas W. Blakiston, *Five Months on the Yang-Tsze* (John Murray, 1862). © Staatliche Kunstsammlungen Dresden, Museum für Völkerkunde Dresden / Eickstedt, Egon von. All rights reserved. https://digital.staatsbibliothek-berlin.de/werkansicht?PPN=PPN618553274&PHYSID=PHYS_0010&DMDID=DMDLOG_0001&view=picture-download.

FIGURE 6.6: An upper Yangzi boat, ready to sail. Source: Alexander Hosie, *Three Years in Western China* (George Philip and Son, 1897), frontispiece.

dominance. As Little noted, the scenery of the Yangzi gorges was "better left unvisited, if it has to be rushed through the stream, leaving no time to study the details or to fix any one picture." At the gorges where silence was complete, he was reminded of Schiller's verses: "Bin ich den wirklich allein, in deinen Armen, Natur!" (Am I really alone in your arms, nature!).[92] Looking at the "stern grandeur" of the Lu-kan Gorge (Figure 6.5), Blakiston invoked Alexander von Humboldt to suggest that "such recollections, like the memory of the sublimest works of poetry and the arts, leave an impression which is never to be effaced."[93] Several decades later, during his trip to the Yangzi, William Dunlop had a similar experience, noting that the "awful panorama unfolded to the lonely wanderer by the God of Nature, in the gorges of the Yang-tse, far in the interior of China, will never fade, or be dimmed, or become effaced, from my memory."[94]

The mnemonic practices of the Yangzi's boat crews and the bewilderment of Western travelers at the sight of the gorges may appear as disparate human experiences of nature, yet a subtle existential cord bound them together. Both the local crew and foreign travelers passed through an utterly challenging water space to move from one region to another, and in doing so, both encountered phenomena that would stay with them. In this time and space, the mundane commuters dwelled on the Yangzi as much as the Yangzi dwelled in them. This co-constitution of interdwelling is at the heart of the commons through which the Yangzi revealed itself.

Once the hurdles of the rapids through the gorges were overcome by mammoth physical efforts, another world of entanglements awaited in Chongqing. The Yangzi provided universalist aspirations, partly reflected in the British imperial quest to connect China and India. As Margary wrote to his parents: "You must picture me standing on the heights of the Momein pass [Tengchong], far away on the Burmese frontier, and anxiously scanning the country beyond for the first glimpse of Indian helmets approaching from the West. Then you can picture the meeting, China and India grasping hands, and awakening those primeval echoes with a British hurrah over the fait accompli." With this, Margary hoped to enjoy the "privileges of doing some service to the world at large."[95] Chongqing, it appears, liberally nurtured this quest.

FIGURE 6.7: Tschungking (Chongqing) am mittleren Jangtse [Chongqing on the middle Yangzi] (hinten rechts; jenseits die Stadt Kiangpe). Drachenbootfest auf dem Jangtse, 1929. Source: © Deutsche Fotothek / H. Spanier. https://www.deutschefotothek.de/documents/obj/90076799.

By the turn of the twentieth century, as Morrison notes, a Chongqing high street was dotted with Buddhist temples, Islamic mosques, and Roman Catholic and other Christian churches.[96] Travelers who visited Chongqing or passed through on their way to India or Burma often added extra flavor to this cosmopolitan space. Blakiston noted how eagerly the four Punjabi Muslim crew accompanying him on an upper Yangzi expedition were greeted by local Muslims and became friends, "'chin-chinning" extensively and bringing presents of sweet cakes and chow chow. These crew members, possessing a copy of the Quran, would work together to decipher its meaning; but as they were not well-versed in their common language of Arabic, "they must have required a large amount of faith to give them any idea of the contents."[97] There are other stories of mingling among and across religious, ethnic, mercantile, and other groups, who brought India, China, and Southeast Asia closer to each other for centuries, as Tansen Sen has shown.[98] Such cosmopolitan assemblages continued in many shapes and substances throughout the nineteenth and early twentieth centuries.

Although some scholars consider the Yangzi a quintessential site of the Chinese nation, it was the most international of rivers in Asia due to the presence of several imperial forces and the tide of mobility of people and commodities that characterized the last imperial century. The Chinese aimed to keep as much of the Yangzi as possible out of European control, while the imperial forces attempted to gain access to as many parts of the river as they could. These frenzied contests over the river often led to conflicts and compromises, as well as forced legal regimes.

Amid the evolving engagements with the Yangzi waters among multiple transimperial stakeholders— including the British, French, Americans, Japanese, and Germans—the river's role as a connector to China's southwestern regions of Tibet, Sichuan, and Yunnan became more prominent, especially after the Nanjing treaty of 1860. These developments further enhanced the precolonial connections between India, Burma, and China via the river valleys of the Mekong, Irrawaddy, and Brahmaputra.

With this story of the Yangzi, the narrative of the BISMRY river networks as a figuration of transregional commons comes to an end. However, there is still one missing piece in the Asian topographical puzzle. Flows and mobilities on the navigable parts of the rivers are easy to appreciate, but how can we reconcile the fact that the closer the rivers came to each other in Yunnan, the more unnavigable they became, with the toughest topography and geomorphology standing between them? How can we sustain the argument of connective mobility along the BISMRY network without resolving the issue of terrestrial immobility faced between the rocky upper reaches of these river valleys? I have already suggested in the introduction that a single pack animal holds the key to this missing piece of the puzzle. The next chapter unfolds its story.

SEVEN

THE ORGANIC BRIDGE*
The Mule and the Making of Asia's Largest Transregion

THE YUNNAN MOUNTAIN RANGES curve up the skyline like a fluctuating heart rate. Frank Lenz, a German-American cyclist, had to cross forty-four ranges of varying heights to reach Tengchong near the Burma border from Sichuan—an experience that aptly describes the region as a "sea of rugged mountains."[1] The interplay of mountains and deep, watery ravines made Yunnan seem almost surreal, prompting a European photographer in the early twentieth century to consider it no place at all, but a "place of pure vision."[2]

From the vantage point of the BISMRY rivers, Yunnan evokes the imagery of a lively crossroads, far from desolate. If the rivers were flowing through finger-like mountains running from a human palm—Yunnan would be the palm, as German explorer and geologist Ferdinand von Richthofen described. Upon reaching Yunnan, choices between the Pacific or the Indian Ocean became merely a matter of which river one wanted to take as a route. From the coastal urban centers at the mouths of these rivers—including Dhaka, Chittagong, Yangon, Moulmein, Bangkok, Ho Chi Minh City, Hanoi, Guangzhou, and Shanghai—Yunnan loomed large, one way or another. These rivers spread out like a necklace around Yunnan, offering numerous routes in such a regular fashion that, as Alexander Hosie suggested in his memoir, "to propose one route for the whole country is like advocating some quack medicine for a patient who lies ill with half a dozen ailments."[3]

* This chapter substantially draws from Iftekhar Iqbal, "The Animal Bridge: The Mule and the Making of Asia's Largest Transregion," *Animal History*, DOI: https://doi.org/10.1525/ah.2025.2458476.

Yet, due to the fierce currents and the rugged terrain between the river valleys, the mobility of people and commodities depended on suitable land transport facilities connecting the navigable points of these rivers. This need was fulfilled by pack animals, with the mule proving to be the most efficient and adored. Moving through a "spider's web of routes, radiating out in all directions," the mule took over where the rivers ceased to be navigable and thus became the connectors of different riverfronts.[4] In the process, the mule formed what may be called an "organic bridge" between the rivers across the highest crossroads of Asia. To appreciate the transregional propensity and practices of each of the BISMRY rivers that aided in nurturing a fluvial commons in Asia, it is important to explore the mobility of the mules in their charm and career.

I
The Mule Across the River Fronts

In 1904, M. E. Willoughby, a major in the Bengal Lancers, published a comprehensive report on the mules and ponies in Yunnan, detailing their commercial, military, and social significance. In the report, Willoughby identified several breeding areas of mules, all of which concentrated near the river banks: along the upper northeastern tributaries of the Irrawaddy leading to Bhamo; the region above Kunlon on the Salween; several areas between the upper Mekong and the upper Yangzi Rivers, with the largest concentration on both sides of the Yangzi around Lijiang; and the neighborhoods of Kunming, mostly near the Red River Valley[5] (Figure 7.1). With an inborn familiarity of montane-riverine landscapes, the mules spanned the navigable points of the river network that connected Yunnan with the seas, with the caravans mostly walking along unnavigable parts of a river or its tributaries.[6] As late as the 1940s, there were at least thirteen major caravan routes, joined by numerous smaller routes.

The use of mules for commercial mobility across Yunnan dates back at least two millennia, although it was not until the development of the mining industry in the late Qing Empire that the mule caravan network expanded significantly. Western imperial ambitions for wider reach across the BISMRY rivers gave mule-based mobility further impetus, especially with the opening of Tengchong as an open port in 1902. The number of mules employed on the road between Burma and Yunnan rose from 9,830 in 1902 to 59,577 in 1930, before the global recession affected the trade flow.[7] By 1937, with the Eastern China Sea blockaded by the Japanese, the Irrawaddy route remained China's only door to the outside world, except for Russia, until this route, too,

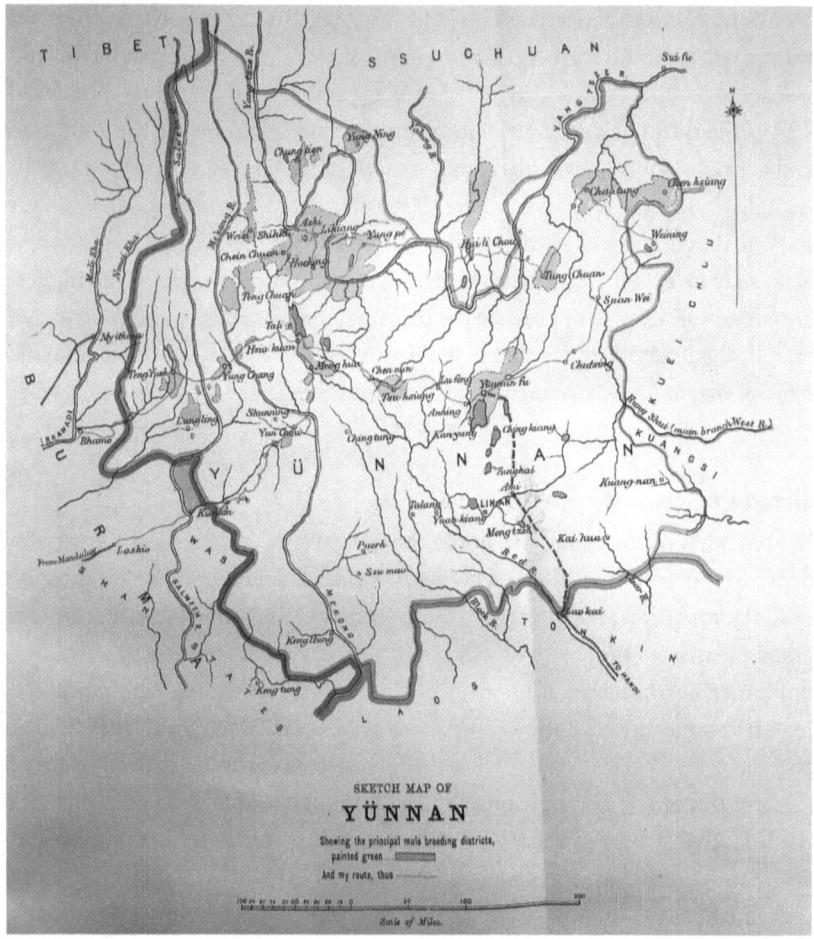

FIGURE 7.1: Major mule breeding areas in Yunnan. Source: The British Library, IOR/L/MIL/7/1061, Collection 14/59.

was blockaded in 1942. These developments led to an increase in mule traffic. In 1941, an average of 150 mules, each carrying 160 lbs. of cotton (about 11 tons in total), cleared through the Tengchong Customs daily and, at the peak of dry season traffic, as many as 500 mules (about 36 tons) entered per day. This excluded a "fair amount" of traffic between the Salween and the Irrawaddy Valley outside of customs radar. British officials expected that with proper coordination between supply chains, muleteer guilds, grazing, breeding, equipment facilities, and Chinese commercial firms, up to 2,000 tons could be carried per month during the war. The cost of mule hiring also rose fifteenfold from the prewar period to 1942–1944, so much so that the prospect of profit induced muleteers to operate caravans year-round, risking the danger of malaria. With the Japanese encroaching from coastal areas all over, the wartime Ledo Road between Assam and Kunming also became partly a mule track.[8] After the fall of Rangoon, the caravans running between the Irrawaddy and Yangzi through Lashio, Bhamo, Myitkyina Baoshan (Yongchang), and Chongqing assumed further importance.

Dali, a key transregional nodal point in Yunnan, was a 40-day mule journey from Yibin, the nearest navigable point on Yangzi in Sichuan; 12 days from Lijiang, which was connected to the commodity flows from Tibet and Sadiya, the last major navigable point on the Brahmaputra; 28 days from Wuzhou, the nearest navigable point on the Canton (Si-Kiang) river in Guanxi; 26 days from Manhao, the nearest navigable point on the Red; and 22 days from Bhamo on the Irrawaddy in upper Burma. Understandably, in Dali caravans of 200 mules "constantly" arrived for tea, with even larger numbers during fairs in April.[9] Lijiang stood on the route from Dali to Lhasa via Batang, where Pu'er tea and other products transited with tea caravans of mules and horses (*mabang*). The existence of at least 17 officially known inns in Lijiang reflected the frequency of caravans transiting from Lijiang to Lhasa.[10]

The caravan connections to the Red River mainly started at Manhao, the last terminal for boats (junk, or *wupan* in Chinese), where Yunnan's products, especially tin, were brought by mules, primarily via Mengzi, and then transshipped to boats. However, a considerable number of caravans moved further down to Lao Cai, following the banks of the Red.[11] Until the construction of the railways in 1910, about 250,000 mules and horses passed the customs station at Mengzi, carrying both import and export items.[12] As noted earlier, with the operation of trains from Haiphong to Kunming in the 1910s, the mule and horse traffic reduced considerably but was never fully replaced. With the onset of World War II, mule traffic was revived.

Across the Mekong network, Forbes identified two main routes between Simao and northern Thailand. The westerly route passed through Kengtung to Chiang Mai via Mae Sai, while the easterly route entered Laos's Phong Saly province, going due south across Luang Prabang to cross the Mekong into Thailand in Chiang Khong. Besides these main routes, between Simao and Keng Hung, there were "numerous passes" of which the British were "totally ignorant, and of which they wished to keep them in the dark."[13] A significant portion of this dual caravan flow, traversing the Mekong, entered Nan and continued as far as Uttaradit on the Nan River, reaching the starting point of steam navigation to Bangkok. In other words, this route connected the Mekong with the Chao Phraya river system.

Routes along the Salween basin generally started at Kentung and reached Moulmein via Chiang Mai. With Hui muleteers charging less, transport and food costs in southwest Yunnan were cheaper than in other parts of Indochina. The cost of transporting goods from Hong Kong through Tonkin and Mengzi was $100 per ton, plus customs dues in Tonkin, while from Rangoon to Simao the cost was $85–90 with no customs dues imposed—making the latter route more popular.[14] The Hui muleteers brought fur-lined coats, straw hats, and copper pots, which were mostly sold in the Shan States. From there, they went to Mandalay to sell the surplus mules and loaded the remaining mules with silk, cotton, and miscellaneous goods to resell in the Shan States on their way back home. The roads on the east bank of the Salween were mountainous, but water, grass, and secure resting places were easily available.[15]

An extension of the Bhamo-Dali mule route, taking twelve days, connected Chongqing via the Jinsha River—a route that had been in place since the Tang Dynasty.[16] Other routes included the Kunlon Ferry-Mandalay Route; and Takaw Ferry-Mone-Hlaindet route (between the Sittang and Salween basins).[17] On the Irrawaddy network, in the 1920s and the 1930s, Bhamo saw a caravan half a mile long arriving from different parts of Yunnan.[18]

In bridging the river networks, mule caravans were driven as much by the desire to connect to distant places as by the profit margins that this mobility afforded. Hallett noted that a caravan of 60–70 mules carried commodities worth at least $12,000 to $15,000, securing a substantial profit margin. For instance, a caravan headman told Hallett that a bundle of 120 straw hats cost him 250 rupees in China and 450–500 rupees in Kiang Tung and even more further down in Bangkok or Moulmein.[19] As the caravans spanned Sichuan, Guizhou, Guanxi, Tongkin, Thailand, Burma, Laos, northern Assam, and Tibet, extending to the Indian Ocean and China Seas, the aggregate profit must have been significant.[20] Consequently, goods imported by the Hui reached the lowland urban centers of Assam, Bangkok, Rangoon, and Hanoi, with Moulmein being reached directly.

II
The Empire and the Ecology of Mule Mobility

The British and the French push for imperial expansion across the BISMRY basins coincided with efforts to expand the railway. However, these initiatives yielded little fruit beyond a bulk of survey reports and travelogues. The primary challenges were the undulating landscape, extreme heights, and inadequate finance. Additionally, the predominance of the river-mule network was strong enough to discourage railway development. In the late nineteenth century, railway connections were proposed between the last navigable points of the BISMRY rivers, including Bhamo, Lao Cai, and Yichang, but such projects fell through as rivers and mule caravans effectively fulfilled the railway's role.[21]

The French Haiphong-Kunming railways, which came into operation in the 1910s, made "little or no difference to the trade carried on by mule-back with Burma."[22] During World War I, when railway construction was particularly emphasized, Kingdon Ward noted that rather than abandoning these "obsolete contrivances," the war created more demand for mules, along with elephants.[23] Even on the eve of World War II, there were no intra-Asian railway networks across southwestern China, Burma, Thailand, India, and Indochina.[24] The railway was never a threat to the mule caravans, not merely because the railway network was limited, but also because in cases where railways did operate, cargo costs were not competitive enough to replace the mule traffic. As Kuo Tsung-fei noted, due to proximity, the cost of overland importation of cotton from India with pack animal was as cheap as from Osaka to Shanghai via the Indochina railway.[25]

Mules in the BISMRY transregion didn't always need to travel all the way to the coastal ports, similar to mule caravans arriving at Genoa, Tripoli, Tunis, or Algiers on the Mediterranean. It was more like the caravans arriving at Verona where the Adige became navigable, with the mule and river traffic conveniently complementing each other.[26] Within the BISMRY transregion, mostly in the case of Moulmein did the Hui caravans proceed as far as the coast. Typically, the caravans followed the tracts along the river bank, with the main caravans often joined by smaller caravans from different interior routes, resembling "incessantly convergent tributaries" joining a major river.[27] With its flow and fluidity, the mule caravan became a metonym for the river in the areas where they ceased to be navigable.

The terrains covered by the mules were rugged, and some European travelers complained that not a single mile was without considerable ups and downs. In some places, stony roads were worn out and lacked grips, so that mules and men "tumbled constantly." Yet, what appeared to be a traveling nightmare for Europeans was a regular encounter for the locals, as British traveler James Turner noted. He observed that the sandals made of bamboo-strings gave the local coolies a sort of precarious foothold, but for Europeans, it was a "serious business to keep one's feet."[28]

This relative ease of mobility was not solely a reflection of the experience and skill of the mules and muleteers, but also of the deep terrestrial affinity they were trained to develop as they navigated the terrain. Along the tough upper Mekong and Salween landscapes, with high mountains and deep valleys in quick succession, mules and muleteers mastered such efficiency that they considered any slope less than 45 degrees as level going.[29] When d'Orléans was wary of passing through a narrow passage on the way between Manhao and Simao, the leader of the muleteers assured him, "if mules can pass, so can others; forward."[30] Not all roads were precarious, though. For example, the roads from Yunnan to the south including Muong-Lu, Luang Prabang, or Lao Cai, were about two meters in width, paved with stone, and had access to cemented drinking troughs for pack animals. This reflected the continuity of historical mobility on well-maintained infrastructure.[31]

In terms of speed, caravans were considerably slower compared to the nineteenth-century oceanic ships. In 1900, a ship sailing from London would run at about 15 knots per hour (17 miles), making it six times faster than the mule.[32] However, as Turner explained, in terms of absolute spatial crossings, considering the mountainous and rugged landscape, the speed of mules and ships would be at par. He noted:

> Chinese count their distances by li, but the li varies according to the nature of the road to be gone over. Three li ought to equal an English mile, and ten li is an hour's journey. So far, as distance is concerned the li is quite unreliable, but in point of time it is very accurate. If the coolies tell you that the day's stage is 90 li you may be sure it will take you nine hours to get over it, although the actual distance instead of being 30 miles may not be much more than a dozen. Only, the 90 li may have taken you 1500 feet down hill, up again 2000, a further descent of 2000 feet (in two miles), and a stiff climb of 3000 feet to finish up with. This is not an imaginary case but an actual day's march.[33]

The relative ease with which the mule caravans operated in the montane zones also made the mules and muleteers vulnerable to and carriers of diseases. The protracted Hui rebellion and the devastating bubonic plague that emerged in its wake in Yunnan significantly decreased the population and hindered commercial growth. However, as noted elsewhere, the recovery of commercial activities and mobilities across Yunnan was visible by the late 1870s, although the scar of the war and the lack of physical well-being continued to haunt Yunnan for a long time.[34]

III
The Interspecies Materiality

The dominance of mules in the BISMRY transregion can only partly be explained by the impenetrability of modern transport. One obvious reason for the mule's popularity was its ability to make long-distance travel carrying heavy loads—up to 215 pounds, which was disproportionate to its own body weight and size. However, the mule's ability to bear loads was only truly advantageous in conjunction with its agility to navigate various types of tracks, making it comparable to a four-wheel drive at its best. A contemporary observer noted the stability of the loads on the back of the mule, which hardly ever made "a false step, safely carrying his load up and down narrow slippery mountain tracks, along stream beds full of loose stones and boulders, through thick greasy mud, through deep water."[35]

Mules also enjoyed a comparative advantage over other pack animals. Discussing yaks and ponies, Kaulback noted that "short of being caught in an avalanche or other act of God, no one has ever seen a yak fall, whether on ice or anything else. The ponies, so sure-footed on rock, found it a different matter."[36] Yet, compared to other pack animals, a greater combination of physical and behavioral attributes made the mule an even more competent companion for long-distance traders. Elephants, for example, were excellent for transporting goods but "too slow, too expensive, and too cumbersome for the distance and the terrain," Forbes noted. Oxen, too, were considered slow and lacking the sure-footedness on mountain trails. Pound for pound, mules were "capable of carrying greater burdens than either oxen or elephants, and over much greater distances."[37] Ponies were better suited to the landscapes than elephants and oxen, but they could not match mules in terms of their ability to cope with both freezing cold and high temperatures, as well as their amenability to training for a calming effect.[38] Willoughby used a caravan of seven mules to travel from Tengchong across Yunnan into Guizhou and back to Kunlon ferry crossing (some 1,600 miles) during the freezing December through April 1904, and they were in "fair condition at the end."[39]

Other skills that distinguished mules from all other pack animals included their ability to climb and descend mountain tracks, packed and with confidence. In his two-and-a-half-day journey from Manhao to Mengzi, d'Orléans noted how his caravan of eighteen mules ascended from 510 feet to 6,150 feet in a short distance, "sturdily in single file, urged by the shouts and imprecations of the drivers."[40] He also found mules to be good mountaineers without suffering from acrophobia.[41]

Despite their smaller size, averaging a height of forty-six inches, and nervousness of strangers, the mules possessed the quality of "endurance, surefootedness, docility, and intelligence" and easily trained to follow the instructions of the *mafus* (muleteer). They were also inclined to swim rivers readily (Figure 7.2), and were accustomed to jumping into and out of ferry boats, with the gunwales of the larger boats on the big rivers sometimes as high as the point of the mule's shoulder, necessitating a very high standing leap (Figure 5.2). Additionally, they went "readily over swaying suspension bridges and frail-looking

Passage of the Lysiang-Kiang, or Black River.

FIGURE 7.2: Men and mules crossing the Black River. Source: Prince Henri d'Orléans, *From Tonkin to India by the Sources of the Irawadi, January '95–January '96* (Methuen & Co, 1898), 72.

bamboo bridges on piles."⁴² This promptness and ambient alertness were informed by their ability to receive human communications. Each *mafus* used personalized calls and sounds—similar to customized ring tones— to draw the mules closer and ensure they followed instructions.

At first glance, mule carriage appears more expensive than bullock carriage; with a mule being paid at the rate of 1.8 rupees to the bullock's 8 annas (16 annas equaled 1 rupee). However, the advantages of using mules justified the higher cost on many counts. Different types of baggage could be loaded on mules, whereas bullocks could not carry anything that did not fit in their baskets. A mule also carried much more than a bullock and could complete a 12- to 15-mile march in less than half the time it would take a bullock. In terms of feeding, a spell of saddle-free grazing and two feeds of bean meal or one feed of rice would suffice for the day. Therefore, mule carriage was at least as cheap, if not cheaper, than bullock hire, and the gain in comfort was "enormous."[43] Another explanation for the mule's popularity is provided by Nani Kim, who suggests that with steeper slopes being degraded and grasslands replaced by farmland, the logic of relying on mules became apparent in the late Qing era, as they had lesser demand for fodder than horses.[44]

The sustainability of the mule-based mobility largely depended on the physical well-being of the mules and their bond with the muleteers. The food and feeding routines of the mules were important aspects of their well-being. As Willoughby noted, mules were given a midday break for about two hours at a convenient place. Their diet included green grass, wheat, green peas or beans, and stalks. Evening meals often included gruel made of flour in chilled water. While the level of care for the mules varied, the freedom to graze during the midday halt and throughout the night was universally provided. Periodic rest between commutes typically occurred after every 15–18 miles and sometimes after 24–25 miles.[45] A longer recess came in the rainy season, following a continuous nine-month spell of hard work that often resulted in injuries to their backs. The extended break was also due to the precarity of the mule tracks and the abundance of grass at home ground.[46] The traditional practice of caring for the mules was officially acknowledged by the colonial administration during World War II. As the road between Myitkyina and Tengchong was being constructed, it was mandated that mules receive proper feeding arrangements and "full protection," and a capping of loading to not more than 160 pounds.[47]

A notable aspect of mule care was the provision of shelter provided to them at various stages of caravan routes. Along the Bhamo-Dali highway there were inns (*tien tzu*), horse inns (*ma tien*), and occasionally official rest houses at regular intervals. The caravanners also had the opportunity to stay at village households near the highways. The cost of accommodation and food for both muleteers (salt, sugar, beef, and mutton) and mules (paddy, rice straw, broad beans, and maize) was affordable. Particularly in the dry season of February to April, when the land was bare and forage was needed most, food was available at these inns at about 3 annas (approximately a fifth of a rupee) per head per night.[48] In some places, facilities for mules were secured even if facilities for muleteers or other travelers were inadequate. Along the roads between Mengzi and Manhao, an English traveler noted the absence of proper accommodation facilities for humans, but found large stables for mules at the recognized halting places, ranging from professionally-managed inns to simple huts. (For example, see Figure 7.3.)

The extent of care for the mules was tested when many of them developed sore or galled backs following continuous service on the road. This problem

Inn between Mongtse and Manhao.

FIGURE 7.3: Men and mules checking in at a resthouse on the Mengtze-Manhao route. Source: Prince Henri d'Orléans, *From Tonkin to India by the Sources of the Irawadi, January '95–January '96* (Methuen & Co, 1898), 38.

was often unavoidable, but the muleteers' long experience with such instances led them to take both preventive and adaptive measures. Young, small, or weaker mules were given lighter loads such as the muleteer's kit. Lighter loads were also considered during hot or wet weather. Both Chinese and Tibetan pack saddles were designed to minimize the friction of loads with the mule's back: "a light wooden frame formed to the curve of a mule's back and had a raised arch in the centre to prevent it from resting on the animal's spine and thus giving it a sore back. Saddles and packs are securely fastened to each other, and are loaded and unloaded together"[49] (Figure 7.4).

A great advantage of this saddle was the possibility of curing the mule's sore back while on the move. The unloading system, which took less than a minute, also helped save both the goods and the mule in times of danger. d'Orléans notes one incident in which a mule's load was quickly off-loaded as it tumbled into a river. The mule then saved itself by swimming across the river before rejoining the caravan.[50]

If a case of galled back occurred despite precautions, it was treated using local remedies and traditional practices. In most instances, the sore back was washed with soap and water, the matted hair around it was cut, and raw poppy juice or vegetable oil was applied. To prevent further rubbing, a hole was cut in the pad to keep the galled back safe.[51] d'Orléans observed that when a mule staggered while on service, Makotou, a head muleteer, bled it from the tongue and burnt a rag under its nose, which caused a discharge from its nostrils. He then "made the animal inhale some powdered pimento placed on glowing charcoal, and finally forced it to swallow a black drug called kouizen. After which attentions the mule revived sufficiently to proceed."[52] Such traditional practices persisted as late as World War II, when the British government found that veterinary services were not required on the Burma-Yunnan routes as muleteers preferred their "own crude but effective methods."[53]

Were the examples of human bonding and care for the mules a reflection of an affective response to the need for cohabiting in a difficult terrain, or were they merely a protocol of convenience in commercial mobility? These questions aside, the mule's affinity with montane ecology, along with their skills, temperament, and tenacity, imbued them with an agency that secured their place among humans and vice-versa, blurring the boundary between "beasts of burden" and the "companion animals," as James Hevia noted in a fascinating study.[54] The human-mule bonding contributed to the growth of cocreated

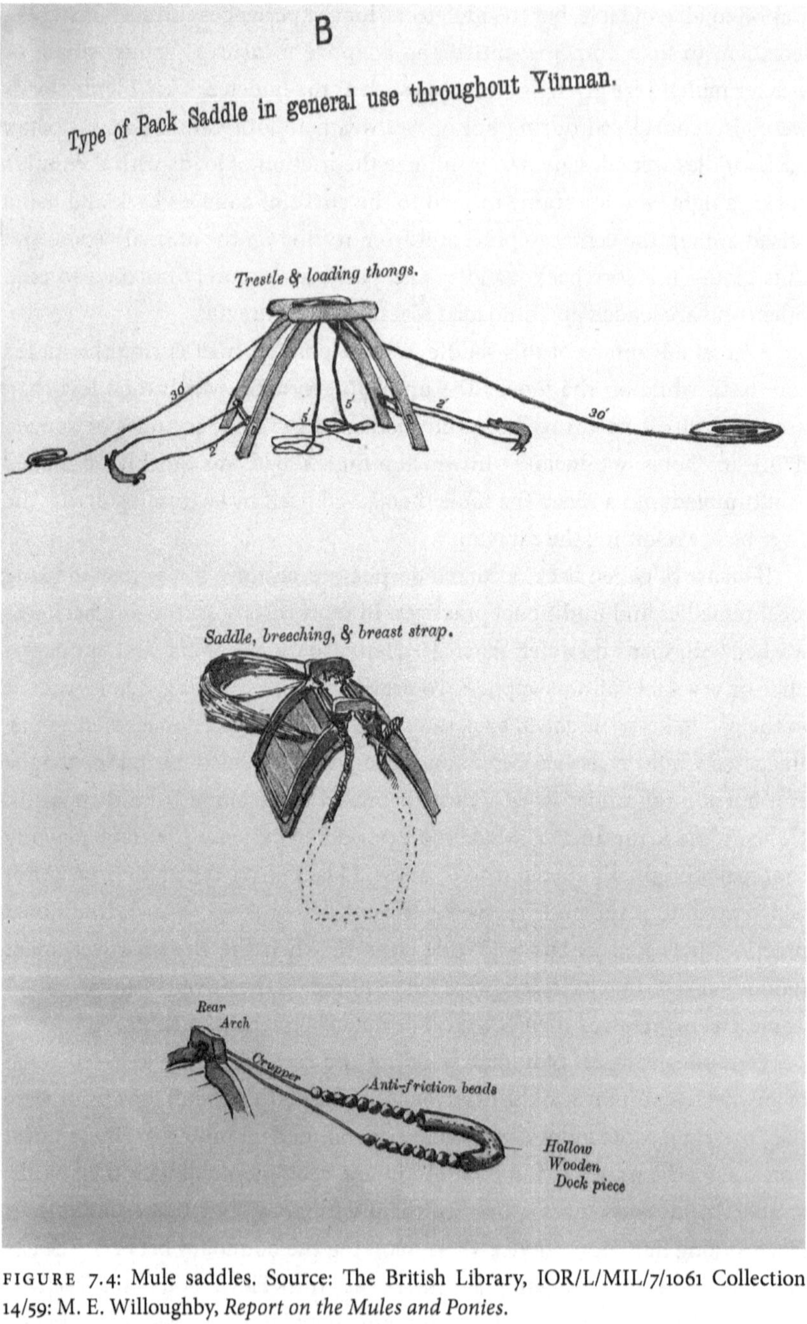

FIGURE 7.4: Mule saddles. Source: The British Library, IOR/L/MIL/7/1061 Collection 14/59: M. E. Willoughby, *Report on the Mules and Ponies*.

intimate spaces as they moved through communities inhabiting the BISMRY river networks—a theme explored below.

IV
The Human Networks Across Caravan Trails

The mules facilitated transregional connections across the navigable nodal points of the BISMRY rivers, but these connections were only possible through the active entanglements of the inhabitants along their paths. As discussed in the previous chapters, the river basins and intervalley regions were inhabited by many ethnic groups with diverse economic and cultural practices. Within this context, caravan managers and the muleteers had to establish social and economic networks. Instead of merely following a transport protocol, the mule caravans forged a crucial relationship between the muleteers and the people of the territories they traversed.[55]

The caravan muleteers were predominantly Yunnanese Muslims (Hui), with a smaller contingent of ethnic Chinese from Yunnan. The commercial exchanges, shared shelter and protection, and social interactions between the Hui and various ethnic groups along the routes nurtured a sustainable pluralism. In Dali, except during the tumultuous years of rebellion, there was a remarkable mingling of Lamas from Tibet, ethnic minorities from Wei-si, Burmese, Cochin Chinese, and people from all neighboring territories, including eleven out of eighteen Chinese provinces.[56] A similar intermixing occurred in Kunming where "Mohammedans, Chinese, Shans, and Lolos, and mixtures of these races, jostle each other in the marketplace and in the daily business of the world."[57] These cosmopolitan feels were primarily brought about by the mobility of diverse ethnic communities facilitated by the mule and muleteers of Yunnan.

As Ann Hill has suggested, rather than an uneven inclination toward commercial relations among the ethnic groups, there were "structural" features of long-distance, translocal trade, primarily shaped by the Hui caravanners.[58] Beyond Yunnan's urban centers, numerous strategic points along the caravan routes saw the emergence of Hui settlements. Those not directly involved in caravanning engaged in service sectors relating to caravanserai, halal restaurants, butchers, and related businesses.[59] Along the caravan pathways also came imams and Sufis, evidenced by the mosques and Sufi settlements between Mojiang, Kengtung, and Chiang Mai.[60] Besides bringing their family members, muleteers negotiated long-distance trading mobility through marriage or membership in local elite groups or secret societies along the routes.[61]

Oral historical narratives from the Hmong living in Laos and Vietnam recount how their ancestors traveled with Hui caravanners in the late nineteenth century to find a new home away from home. They worked as grooms for the horses and mules of the Hui caravanners and moved south with them for a living. As Michaud and Culas demonstrate, the availability of fertile and forest land near Hui caravan routes was ideally suited for the moving Hmong, who sought to "escape the Han wrath and to try their luck."[62]

With the Akhas in northern Indochina, the relationship ranged from marriages with local women to direct commercial transactions, involving silk thread, gold, and silver in exchange for Akha rice, wine, and white cotton.[63] In addition to trading at the well-known fluvial nodal zones of Luang Prabang, Moulmein, and Chiang Mai, the Hui traded in the "forbidden" Wa territories in precious stones, jade, opium, and copper pans, among other items.[64]

As Forbes has noted, major Hui settlements emerged in Burma following the failure of their rebellion, notably at Panglong, on the east bank of Salween in the Wa States. This was followed by settlements at other points in the Salween basin, such as Tanyan near Lashio, and other major urban centers, including Rangoon, Mandalay, Taunggyi, Bhamo, Mogok, and Kengtung in the Shan States.[65] Many settled just above the Salween Valley near Tangyan, contributing to the trade network, which was replete with tensions and competition, but was characterized by "continuous circulations of a wide range of objects."[66]

In the Bokeo region in Laos, particularly in the Lamet territories on the Nam Tha River, a long tributary of the Mekong, the connections between the Hui and local ethnic groups were firmly established along the mule track. An old caravan path traversed the whole province as far as Kunming, serving as the main line of communication for the region. The villagers set up *cong ying* or community houses, for the caravanners.[67] This bonding with the Hui connected the Lamets to a wider world. As Izikowitz noted, the steel used by Laotian blacksmiths was of Swedish origin, and consequently, the iron implements of the Lamet were made of Swedish steel. He remarked, "It is quite strange to see how far world commerce stretches its tentacles into the jungles."[68]

The Kachins were both buyers and toll collectors from the mule caravans, a fact that shaped their dynamic relationship with the Hui. The tolls they extracted were reasonable enough to keep the route profitable for the Hui, and the Kachins provided security for the caravans since part of their livelihood depended on the caravans passing through their territories. Any deviation from this norm could have driven the Hui away from the Kachin routes.[69]

Similar networking developed between the Hui and other ethnic groups along the caravan routes, including the Yao, Miao, and Lisu, and Lahu, Lolo, Dai, and other upland people in northern Thailand. For those groups, the Hui were an "accepted part of everyday social environment," and some ethnic groups spoke in the Yunnanese dialect of Mandarin.[70] The Hui relationship with the Shans was perhaps the most extensive, as their trade routes crossed through Shan territories, especially along the Salween River.[71] It is these Shan-Hui caravan connections that also facilitated the Assamese merchant's outreach to the Yangzi Valley.[72]

Chiang Saen, situated on the plains on both sides of the Mekong, was a crossroads of routes from China, Burma, Karenni, the Shan States, Siam, Tonkin, and Annam, serving as a "centre of intercourse between all Indo-Chinese races."[73] To the south toward Thailand, the Nan River, originating in Chiang Rai, was navigable from Uttaradit to Bangkok for large barges that carried commodities from Luang Prabang and markets along the Mekong and vice versa, as discussed in chapter 4. The transshipment between these lands and navigable waterfronts was facilitated by mules and oxen. Since the Salween was unnavigable until almost to the sea, the mule track was extended to reach the coast. The presence of the Hui along the Salween River, principally in Kengtung, Mandalay, Taunggi, and surrounding areas, reflected their growing population, which reached at least 15,000 at the beginning of World War II. Another wave of Hui settlements occurred after the Communist takeover of China. The late twentieth-century Hui settlements in this region represent a continuation of the historical mobility along the middle Mekong to the lower Salween.[74]

The history of interactions between Hui caravanners and the ethnic groups along the caravan routes reveals how translocal collaborations at major navigable nodal points of each of the BISMRY rivers defied geomorphological divides and achieved what empires could not. The Mishmi around Sadiya, Kachins around Bhamo, Shans, Lolos, and other groups in the upper Salween and Mekong regions around Luang Prabang, along with the Black Flags and Hmongs/Mias around Lao Cai, all demonstrated a common pattern of guardianship over the transitional zones between the plains and the highland river valleys.

The gap between the quest for mobility and the challenging natural settings in Yunnan was bridged through intensive interactions among the merchants, ethnic or religious groups, and other human actors such as smugglers and secret societies, effectively dismantling spatial segmentations. These networks were cemented by the Hui mules and the caravan system. It is no surprise that the Hui concentrated in the transportation hubs along the rivers to facilitate trade and integrate with local population through social interactions, marriage, and conversion. Moreover, Hui leaders regularly explored the rivers and valleys to overcome topographical difficulties and develop new irrigation systems that benefited the communities they interacted with.[75]

The business of caravanning was indeed impacted by political upheavals, natural disasters, and bouts of epidemics. Travelers from the Yunnan and Sichuan uplands were vulnerable to malaria, as David Bello has rightly noted. However, recent studies also demonstrate that these barriers were not insurmountable. During the Hui rebellion, traffic was disrupted, but never stopped and found alternative ways. Huis made their downward journeys during the "least unwholesome season and acquired skills of dealing with illness. Although boundaries existed, they were permeable and extended by the routes of the muleteers."[76] The expression *xia jiangbianpo*, meaning "to walk downwards to a river," reflects a cultural embeddedness to a basin-based mobility that was epitomized by the mule caravans.

In a tribute to Fernand Braudel, John Leonard quips that he revealed how most history is written at the expense of ordinary people—"maybe with hollow-bladed scissors, to cut us down to size so that great men will have somewhere to sit or stand." But, as Leonard notes, Braudel's perspective, "heroically, is that of the pack animal."[77] In the BISMRY transregion, the mule network adapted to the heightened transregional flows driven by transimperial mobilities, while preserving its autonomy due to its deep acquaintance with the local ecological conditions.[78] The mutual human-mule dependency that developed out of this mundane necessity was at the heart of what Donna Haraway calls the Chthulucene, signaling "multispecies stories and practices" where both human and nonhuman actors created a kinship that fostered a sort of "response-ability."[79] Beyond Haraway's ahistorical conceptualization of human-animal bonding, Jocelyne Porcher proposes that it is specific "work" protocols that shape the intersubjective bonding between human and animal while reflecting "their resistance and their propositions."[80]

In this broader context of interactive and responsive kinship, developed through embodied labor of both humans and animals, the mule becomes central to the story of *The Range of the River*. Yunnan had the Irrawaddy, Salween, Mekong, Red, and Yangzi cutting through its territories while the Brahmaputra network reached its borders. Despite the outreach of these rivers, the rugged and fluctuating terrains between them required the services of a seasoned transporter to ensure the continuity of flows between navigable parts of the rivers and their destinations. In this context, if Yunnan was the "sea of rugged mountains," mules were the ships on this sea, and the muleteers were the sailors.[81] Thanks to the mules, the upper BISMRY network in Yunnan was not a chokepoint but the jugular vein of Asia's largest and highest crossroads, serving as a common cradle of flowing sweet waters.

REFLECTIONS

"OXEN IN A FLOWER GARDEN"
The Postcolonial State and the Fate of the BISMRY Commons

THE RANGE OF THE BISMRY rivers explored in this book represents a space as vast as it is ecologically diverse, marked by equally intricate entanglements between human and other-than-human worlds. These rivers spanned South Asia, mainland Southeast Asia, and southern China via interactions between high powers, riparian ethnic communities, long- and short-distance traders, and pack animals. All these actors competed for a space in the river networks, but none were able to exercise exclusive domination due to uncertain political conditions and terrestrial difficulties. This interplay created a foundation for coexistence, collaboration, and the cocreation of a large-scale fluvial commons. The commons here are not nature's timeless endowment but are cocreated by multiple actors through a range of dynamic frictions.

Bolstered by initial success in securing the coastal ports on the Indo-Pacific rim, external imperial powers sought access to inland territories as far as Yunnan by following the river network. The floodplains in the deltas were the first point of encounter with the empires and received greater shocks, yet some zones in the deltas remained inaccessible even with the aid of steamers and gunboats. Railways, too, could expand only marginally and hardly rivaled the rivers, allowing local inhabitants, boaters, everyday travelers, and merchants to remain the primary architects of mobility along the river network.

As the empire moved up the BISMRY rivers, its authority waned noticeably, almost disappearing at the montane edges where the rivers ceased to be navigable. However, as this book has argued, at the upper riparian "choking" zones where the empire lost its grip, a renewed impetus for transregional assemblages, rather than prescriptive isolation, became the norm. This involved an intimate collaboration among boaters, porters, and muleteers—a multistake entanglement of fluvial vibrancy and human and animal labor at the river sites, partly resembling what Richard White calls the "organic machine."[1]

Each river, with its elemental freshwater regime, fostered assemblages of political forces, interethnic entanglements, and shared spaces of mobility at various fluvial points: the deltaic plain, spaces between the delta and the upper torrent, the confluences, and along the land trails projecting from the highest navigable points toward the navigable zones onto the next river. The continuity of mobility across the uneven landscapes, from the Tibetan-Himalayan highlands to the Indo-Pacific rim, underscores Bruno Latour's conception of a nonlinear spatial scale. In this framework, entanglements between points A and C may be more plausible than between points A and B, as spatial scale is shaped more by place making and networking than terrestrial proximity.[2]

Global commodity flows intercepted at these rivers' sprawling but unnavigable headwaters were rerouted to land-based commercial networks connecting to the navigable points of the next river. As these networked mobilities signaled the fusions of local, regional, and global dynamics, the notion of ethnopolitical autonomy embedded in the folds of the mountains faltered at every river confluence, at the fordable parts of the river, and along the mule tracks between them. In other words, the BISMRY rivers proved to be ruthless levelers, dismantling the boundaries of area or regional studies shaped by structuralist anthropologists and scholars constrained by the logic of Cold War territorial visions in Asia.

The co-creation of the river commons was secured by their role as resilient agents of nature. Due to their sheer physical flow, uncertain and seasonally variable velocity, rapids, falls, rolling stones, silt, mud, bars, and floods, these rivers were forces that did not readily yield to imperial advances. Despite the exploitation of resources taking multiple forms, from deforestation to mining with significant ecological ramifications, the rivers generally escaped being an object of conscious anthropogenic onslaught. Their navigable parts had to be kept in their natural state as essential arteries of trade and transport. Even when deforestation was rampant, the river channels needed to remain free flowing to facilitate the floating of the logs for downstream ports.

The prominence of the BISMRY rivers as transregional and common sites of trading and commercial flows was underscored by the comments of

Nathaniel Curzon, a controversial viceroy of British India and a keen geostrategist in Asia. Pressured by British investors to construct railways across Burma, Thailand, and Yunnan, he disapproved of such a scheme and quipped in 1897:

> The idea that if it [Burma-China railways] were built the wealth of Szechuan would stream down a single metre-gauge line, many miles of which would lie over mountains, to Rangoon, while great arterial rivers flow through the heart of Szechuan itself, which are quite competent to convey its trade to and from the sea, is one which seems to me in the present stage of Central Asian evolution almost of midsummer madness.[3]

The conceptual boundary between the empire and the postcolonial state is certainly porous, as the seeds of unsustainable economic and development projects were sown in the time of modern empires. Indirect effects of deforestation and railway embankments in parts of deltas like the Bengal and Red Rivers began to impact the free flow of the rivers, leading to a disturbed ecosystem and deteriorated public health. Yet, the reality of the BISMRY rivers as commercial highways and their untamable nature preserved much of their properties during most of the imperial period.

On the upper Irrawaddy, for example, early discussions on dams did not materialize. In the early twentieth century, a British colonial official reported that of the Mali and Nmai Rivers that formed the Irrawaddy, locals considered the Mali as a "good [big] river" and the Nmai as a "bad river." The author attributed this distinction to an "Oriental theory," where the "big" river was favored as it was navigable and the "bad" river was characterized by its non-navigability. To him, the river's importance was measured by its energy potential, as reflected in the flow per cubic feet per second.[4] This modern scientific appreciation of the potential of a "good river" as a source of energy rather than a laterally agile, free-flowing body of water, did foreshadow the construction of dams, but actual dam projects that affected the rivers had to wait until much later, in the 1950s.

The imperial era, therefore, generally saw the continuity of the river as a common site of high mobility of ethnic groups and productive activities. One form of mobility involved people adapting to evolving agroecological zones that connected both upstream and downstream areas. Another form spanned far larger spaces along and across the riverine systems, influenced by navigability and commercial networks, often ushering in forms of cosmopolitanism.

As entanglements between the ethnic groups over the uses of rivers opened up new possibilities for human mobility and productivity, a mutually constitutive process developed, which I would call the "ethno-morphology of rivers." This ethno-morphology, an outcome of long and sustained interactions between ethnic groups and diverse riparian ecologies, reflects a material relationality between human and nature that moves beyond the Cartesian dualism and embraces what Bird-David terms "I relate, therefore I am." This relationality is what shaped the contour of the river commons.[5]

The delayed Anthropocene in the BISMRY transregion was partly due to the absence of modern transportation facilities for carrying building materials, the lack of appropriate technologies for negotiating high altitudes, and inadequate finances.[6] Additionally, high-impact development projects were not launched because of the ecologically informed appreciation of the multifaceted benefit of the free-flowing river beyond Curzon's narrow commercial perspective. For example, British officials working in China discouraged modern, sweeping engineering projects as in the form of the Tennessee Valley Authority. They recognized that issues of flooding, irrigation, drainage, energy, trade, shipping, and agriculture were "often likely to prove entirely local, sometimes dependent on each other, sometimes opposed to each other." In this context, it was suggested that any attempt to curb the Yangzi's "freedom" would be "a serious and extensive business" and that the large expenditure required needed to be "carefully weighed against the practical benefits expected therefrom."[7]

If the BISMRY rivers, with all their strengths, were able to assert themselves as a crossroads of multiple forces that shaped a transbasin commons, they have entered a dystopian afterlife in the era of the postcolonial state. With decolonization, rivers as sites of translocal and transregional commons were overshadowed by the securitization of borders and the control of the movement of people across the newly fashioned borderlines. The empire was not entirely successful in undermining the ecological commons; now this task was passed on to the postcolonial state, which reinforced capital- and technology-intensive projects that the terrestrial empire had been unable to implement. Rivers came to be viewed as a "resource" subject to manipulation, existing to serve the nation, while the claims of neighboring riparian states were reduced to mere remnants of the river's full expanse. Decolonization in Asia thus inaugurated the process of decommonization too.[8] While the commons in the BISMRY region had not yet fully manifested as a "tragedy" in the prenational era, they undeniably became one in the postcolonial era.

II

While the importance of this vast water space can be hardly overstated, challenges to its existence as a planetary commons have mounted over the recent decades—largely triggered by national development processes.[9] There are a few terms that have been employed as a trope for advancing respective national interest in the river spaces: "water sharing," "water diplomacy," "integrated water management," and so on. These vocabularies have emerged from the usual practices of transnational water relations in which rivers are seen as a container of water that can be measured in cubic feet per second; then riparian states can set rules, take their share, or negotiate, engage with, follow, or ignore the rules. If nothing else works, the possibility of conflicts and wars is not ruled out.[10]

The vivisection of the BISMRY commons is most glaringly highlighted in the unyielding enthusiasm for the construction of dams. In the river basins spanning India, Bangladesh, Myanmar, Thailand, Cambodia, Laos, Vietnam, and China, there are close to a thousand large hydroelectric dams, excluding several thousand mixed and water supply dams. Whatever the concerns specific to each of the BISMRY rivers, several common challenges are discernable. One relates to the danger of massive humanmade disasters. A vast majority of dams are located on active tectonic fault lines. In recent years, for example, scientists have expressed concerns about the correlation between the high water levels at the Three Gorges Dam reservoirs and seismic disturbances in western Sichuan. The USGS detected a thirtyfold increase in seismicity between the construction of the Three Gorges Dam in 2003 and 2009, leading to more than 34,000 minor earthquakes in addition to numerous landslides. There are ongoing debates about the link between the dam and the 2008 earthquake, which killed about 60,000 people.[11] If the Ataturk Dam, with a reservoir capacity of 48.7 billion cubic meters of water, and the 2023 earthquake in southern Turkey were even marginally linked, a major earthquake affecting the dams on the BISMRY fault line zones could potentially alter the landscape of human and nonhuman habitats across a large swath of fluvial Asia.

A second set of issues relates to agroecological challenges around the shifting scale of sedimentation. Sediment has been a central hydrogeological property in the Tibetan rivers, which have formed and fertilized lands in the delta regions, provided facilities for crop production, and contributed to the growth of mangroves and other forests, as well as biodiversity. Between the 1960s and 2010s, due to the more than two hundred mega-dams and tens of thousands of small and large reservoirs, most of these rivers witnessed more than a 75 percent reduction in sediment transport. This reduction has been proportional to the construction of the dams upstream.[12] A more immediate impact has been the loss of fish habitats and restrictions on fish mobility, which have affected fish productivity and livelihoods. For example, on the Yangzi alone, the fish catch came down to below 100,000 tons in 2010s against 430,000 tons in the 1950s.[13] The potential of silt as a land-forming tool against sea-level rise across the BISMRY coastal zones is also diminished due to the construction of dams and other infrastructure upstream.

In the realm of everyday human interactions, dams have contributed to the gradual invisibility of the river as an inclusive space. The confluences of the BISMRY rivers and their tributaries, once central sites of ethnic settlements and transregional commerce and mobilities, have become the exclusive domain of dams, displacing many ethnic groups. More alarmingly, dams have severed the intersubjective portal for communication between the river and both human and nonhuman actors. The growing dominance of dams over human perception is so overwhelming that, as Andrew Johnson quotes from one of his interlocutors on the Mekong at the Thailand-Laos border: "At night, I close my eyes and all I can see is the dam."[14] Where nature and humans once dwelled together, that possibility of interdwelling has been thoroughly displaced by the dams.

Another major challenge facing the BISMRY commons relates to schemes for the artificial diversion of their waters. The recent Chinese and Indian efforts to harness the Tibetan water regime may be considered a late episode of the Anthropocene that dwarfs European and American hydrological projects. The Indian government has begun implementing a $250 billion "National River Linking Project," aiming to divert waters from the Ganga and the Brahmaputra to its drought-stricken northwestern and southern regions. China's reach is even longer and bolder, having initiated its massive "South-North Water Diversion Project" to divert 45 billion cubic meters of water per year from the Tsangpo, Yangzi, and Han to quench its dry north for $63 billion.[15] These examples of manipulating rivers since the beginning of the twenty-first century, dubbed "zombie projects" by a commentator, invoked the specter of Wittfogel on a larger scale, a specter that had spared the region until recently.

A recent development that has overshadowed the river as a commons relates to the transregional economic schemes of India and China. Since 2013, the $1 trillion Belt and Road Initiative (BRI) of China seeks to revitalize the ancient land and maritime Silk Route that spans the continents of Asia, Europe, and Africa. Fancy bridges are fast spanning these rivers, speed trains are replacing the mule tracks along them, and electricity is being generated from the strangled currents of the river that will lead to the production of commodities that travel through the BRI network. One example is that the high-speed railway line from Yichang to Wanzhou, opened in 2010 after seven years of construction, has been the "most expensive and difficult railway ever built in China." With 74 percent of the 377 km railway line running through tunnels and bridges, it appears that rivers have finally been subdued.[16] India's transregional outreach is not as pronounced as the BRI's, but its "Cotton Route" across the Indian Ocean is claimed to have rivaled the Silk Route in the past and is capable of doing so again.[17] India's Look East Policy toward Southeast Asia complements this reenactment of historical connectivity. But, like in the case of the BRI, such megaregional protocol of trade routes, where Indian commodities will flow, are integral to the production of energy and commercial crops by damming and diverting the Brahmaputra and other northeastern Indian rivers. In the new display of mobility and connections originating from two rising economic giants of Asia, the BISMRY rivers have been relegated to the back burner of these growth trajectories from being the historic shared sites of economic activities.

These challenges have been exacerbated by the far-reaching impact of climate change, which is set to fundamentally alter the BISMRY commons. Since the middle of the twentieth century, the Asia-Pacific coastline has seen the sea level rise between 3 to 4 mm per year. This has increased salinity in the freshwater supply of all the BISMRY rivers. Additionally, the sluggish currents caused by upstream dams further aggravate these problems. The changes in the BISMRY delta regions due to global warming have further worsened the situation, diminishing the ecosystem services of these rivers, and affecting the summer rice cultivation and other economic and livelihood options. What historians once saw as another Mediterranean—a stable seascape with great commercial and cultural vibrancy—now faces a bleak future.

At the other end of the spectrum, near the place of origin of these rivers in the Tibetan-Himalayan upland and Southeast Asian massif—which many anthropologists found to be an unspoiled and autonomous abode of human and nonhuman actors— climate change has also taken a toll. The rainfall in Kunming, around which the BISMRY rivers flow, has decreased by more than 100 mm over the last century and a half. Rising temperatures, such as a high of more than 45°C in parts of the BISMRY region, threaten faster glacial melting, leading to flash floods and the risk of dam collapses in the medium term, and the eventual death of snow-fed rivers in the long run. Pollution and sooty smoke from South Asia have also contributed to the retreat of the Tibetan glaciers. Unsurprisingly, recent data shows that the Tibetan Plateau has lost over 10 billion tons of water a year since 2002.[18] By the end of the twenty-first century, it is conservatively estimated that at least 84 percent of the Tibetan Plateau faces terrestrial water shortages.[19] Besides the decline of the terrestrial base of life and livelihood, climate change has given rise to adaption measures that, while well intentioned, are set to erode the memory of intimate human-river entanglements as portrayed in this book.[20]

The fact that these rivers were once accessible commons that transcended constructed political boundaries, creating an unregimented space of mobility and well-being for both humans and nonhumans, has largely been forgotten. This amnesia about Asian fluvial commons is catastrophic, especially at a time when we needed to engage the collective ecological power of the river to confront the impacts of anthropogenic climate change and the effects of El Niño and La Niña across the region. These rivers are essential for feeding and nourishing billions of human and nonhuman lives, and for sustaining the creative imagination and cultural practices centered on fresh water.

III

Specifically, in the case of the Brahmaputra, independent India adopted the late-colonial notion that excess water was flowing into East Pakistan (now Bangladesh since 1971), which led to the construction of the Farakka Barrage. Just above the confluence of the Ganga-Brahmaputra, a significant portion of Ganga's water was diverted to the Hugli River in West Bengal via the barrage, resulting in an agrarian and ecological decline in Bangladesh and parts of West Bengal itself.[21] The Teesta, a major tributary of the Brahmaputra, is also dammed for irrigation and hydroelectricity within India, which refuses to enter into any agreement with Bangladesh to share the river's water. Numerous other dams have been built or are under construction on many tributaries of the Brahmaputra, which formerly cemented economic and cultural relations between northeastern India and the upper Irrawaddy areas with the

Bengal Delta. Since 1957, dozens of dams, with a height ranging from 24 to 130 meters, have been built in the states of West Bengal, Arunachal Pradesh, Assam, Sikkim, Meghalaya, and Nagaland—choking, reducing, and diverting the flows of the Brahmaputra away from the Bengal Delta. Further upstream, beyond the Indian borders, China has completed similar projects in Tibetan parts of the Brahmaputra (Tsangpo), affecting its flows downstream in India and Bangladesh.[22] Within Bangladesh, the Brahmaputra (Jamuna), which was once a key element of national imagination, has recently seen a government proposal to "straighten" it by reducing its width to 6.5 kilometers from 15 kilometers, with the anticipated support of the World Bank.[23] This is reminiscent of the long-abandoned nineteenth-century project of "rectification" of the Rhine and a few other rivers through straightjacketing that came at the cost of the river's hydrological health, cultural repertoire, and historical footprints.[24]

The confluence of the Mali and Nmai Rivers, where the Irrawaddy is born, has been a vital site for the cultures and livelihoods of numerous ethnic groups and a hub of transregional mobility. This confluence has now become a major site for Myanmar's largest dam projects. The Shweli River was dammed in 2008, and dams are under construction on the Taping River. The largest and most challenging project, the Myitsone Dam, is set to materialize soon.

The political powers that once vied for control of the freely navigable Irrawaddy and its banks for transregional trade have recently collaborated to build dams at this crucial confluence. This effort, known as the "Confluence Region Hydropower Projects," involves at least nine companies from Myanmar, China, and Japan, led by the state-owned China Power Investment Corporation. The project began in 2009 but was halted due to resistance from the Kachin and other affected minorities. Despite the halt, the project has not been abandoned, and those displaced at its onset have yet to return to their homes in these "confluence" zones. The dam is expected to be commissioned sooner or later, thanks to the geostrategic relations between China and the Myanmar government.

In the last century, the discharge at Bhamo has decreased to less than half a million cubic feet per second, compared to the levels recorded in 1907.[25] A major dam at the headwaters would further deteriorate water accessibility in the lower Irrawaddy. During the time covered in this book, rivers and ethnic dwellings strengthened each other, and resistance was a bargaining chip at the site of the river commons. In postcolonial times, political resistance has been about bare survival against state-sponsored encroachment, often mediated by global capital, into the riverine ecology, local livelihood, and cultural identities.[26]

The Salween is often advertised as the last free-flowing river in Asia. However, this is not entirely accurate when considering the state of its tributaries. The Salween's tributaries in China have been extensively dammed, and at least four hydroelectric dams on two tributaries within Myanmar have been completed. Three dams on the Baluchaung River in Karenni State and one on the Teng River in Shan State were constructed between the early 1970s and 2009. Nine more dams are either under construction or in the planning stages on five major tributaries. Preparations for dam construction on the Salween itself are also ongoing.[27]

Similarly, at least 745 dams, including 209 hydroelectric dams, dot the Mekong basin as of 2024.[28] This prompted Brian Eyler to see ominous signs of the death of the river as he traveled from its source to the sea.[29] The Mekong is caught between the idea of economic integration, as promoted by the Mekong Commission, and the unsustainable fragmentation operated by the dams. A recent study suggests that the river has become a central metaphor or a rationalizing image that would propel dam building, hydroelectricity, and ultimately economic development via industrial growth—a circulatory process that occupies a new "epistemic space."[30]

On the Red River, as Li Tana has observed, change has been constant over the past two millennia. However, ecological changes became more pronounced since the mid-nineteenth century when the French coined the name "Red," reflecting the red color resulting from deforestation and erosion of topsoil in the mountains along the Yunnan-Vietnam borders, exacerbated by logging and dyke building.[31] In the late nineteenth century, some observers suggested that the color was due to mining upstream areas. Both explanations are plausible, but the worst came in the postcolonial period when the Red entered the dam regime.

Inside China, the Red River basin now has 41 hydropower dams, 2 multipurpose dams, and 25 irrigation dams. Within Vietnam, there are 25 hydropower dams, 3 multipurpose dams, and 9 irrigation dams. Beyond the usual ecological, agricultural, and social impacts, these dams often cause interstate tensions between China and Vietnam as the latter frequently complains about unannounced water releases from the Chinese dams causing flash floods and resulting calamities. This is similar to the tensions between India and China as well as India and Bangladesh over the sharing of the Ganga and Brahmaputra

Rivers. The concept of fluvial commons is further marginalized with increasing homogenization and market visibility, a phenomenon Sarah Turner calls "infrastructural violence."[32] The Red's sister river, the Black, has met with similar consequences under this dam regime, which dismantles local specificities in the name of national development. Once a roaring connector between upland and plainland, the Black is now a "tamed river, chopped up by dams providing energy."[33]

The Yangzi began to show signs of decline in the late Qing Empire, as chronicled by Yan Gao. However, as late as the early twentieth century, the river's ecology continued to be relatively stable. For instance, Archibald Little noted that the fish population in Sichuan was "small, scarce and dear," but in the Yichang gorge, shoals of porpoises could be "daily seen . . . rich swarms with every description."[34] Drastic changes began in the 1950s when dams started to dominate the Yangzi waterscape, with more than twenty thousand dams over 15 meters in height erected since then.[35] With the construction of the Three Gorges Dam, the decline of the river has reached a full circle. Some of the largest dams have been constructed in Yunnan and Sichuan, especially around the Chongqing fluvial network. As a result of the dams, combined with fluctuating rainfall, pollution, and oversiltation, as many as sixty-six smaller rivers across Chongqing have dried up, according to Chinese media. Ironically, for example, the Three Gorges Dam was unable to produce enough electricity due to drought in August 2022, which resulted in the shutting down of industrial plants in Sichuan for a week. The stunning beauty that once hypnotized travelers has turned into a beast.[36] All of these point to the fact that the vibrant flux of encounters, collaboration, and transregional mobility that once sustained the Yangzi as a commons is now fraught with existential anxieties.

IV

For centuries, the Tibetan rivers evolved as a fluid commons—uniting empires and subalterns, animals and fishes, traders, boaters, and peasants across the regions. The postcolonial state, driven by rapid dam constructions, water diversions, and climate change, has charted a different path for these rivers, compromising their historic role as shared spaces. As we confront the current phase of the Anthropocene, scientific innovations, social movements, and positive policies aimed at preserving our planetary lifeblood may find inspiration in the predevelopmental histories of the BISMRY rivers.

The river's agency was at its best when it assembled conflicting forces and promoted a common ground—a dialectic process powerful enough to delay the full-scale advance of the Anthropocene. As the twenty-first century progresses, we have crossed that threshold, but perhaps not irrevocably. *The Range of the River* aspires to return us to the heart of a fluvial Asia, where an intimate lifeworld once thrived.

Rivers are intimate because their properties are alive and active. The freedom and the mobility of the river—and the fish, silt, and other animate and inanimate beings that move with it across varied topographies—is commensurate with the life-nourishing benefit of the commons they harbor. Reimagining the BISMRY commons will, therefore, require a connected view of the mountains, plains, coast, and mobile river networks that bind them together. The snow on the mountain cap and the silt in the deltas are just one ecological manifestation of this commons. Navigable or not, the Tibetan rivers were historically a colossal artery of mobility, illustrating how ecological, cultural, and economic commons mutually informed each other. Reminding his readers about the potential calamity caused by humans behaving like "oxen in a flower garden," Thoreau once suggested:

> As in many countries precious metals belong to the crown, so here more precious natural objects of rare beauty should belong to the public. Not only the channel, but both banks of every river should be a public highway. It is not the only use of a river, to float on it.[37]

Keeping in mind the lived experiences and entangled histories of the Brahmaputra, Irrawaddy, Salween, Mekong, Red, Yangzi and their neighboring networks, I argue that these rivers challenge the notion of waterways as mere slices of nature existing externally to human lifeworld—ready to be engineered, packed, and delivered into the development regime of the nation-state. The memory of the river as a commons may have faded from our postcolonial consciousness, yet traces remain, waiting to be reclaimed by a new generation of environmental historians. With that hope, I close this book by echoing the beautiful words of Astrida Neimanis:

> As glaciers melt, deltas flood, and we row our lifeboats down the middle of the River Anthropocene, it seems we need any valuable tool we can muster to negotiate the rising tide pushing in from the sea. Bodies of water—as lived embodiment, as figuration, as hydrocommons in difference, and as feminist protest—may not be the paddle that will guide us out of this planetary mess. But I am wagering that this figuration might just help us learn to swim.[38]

NOTES

Introduction

1. Alan Roe and Iftekhar Iqbal, "Riverine Environments," in *A Companion to Global Environmental History*, ed. John R. McNeill and Maulden (Wiley, 2025).

2. David Gilmartin, *Blood and Water: The Indus River Basin in Modern History* (University of California Press, 2015); Ruth Mostern, *The Yellow River: Natural and Unnatural History* (Yale University Press, 2021); Ling Zhang, *The River, the Plain and the State: An Environmental Drama in Northern Song China* (Cambridge University Press, 2016); Sudipta Sen, *Ganges The Many Pasts of an Indian River* (Yale University Press, 2019); Arupjyoti Saikia, *The Unquiet River: A Biography of the Brahmaputra* (Oxford University Press, 2019); Dilip da Cunha, *The Invention of Rivers: Alexander's Eye and Ganga's Descent* (University of Pennsylvania Press, 2019); James C. Scott, *In Praise of Floods: The Untamed River and the Life It Brings* (Yale University Press, 2025).

3. F. v. Richthofen, "Recent Attempts to Find a Direct Trade-road to South-western China," in *Ocean Highways: The Geographical Record*, ed. C. R. Markham (London, 1874), 404.

4. For recent debates on these hydrological shifts, see M. K. Clark et al., "Surface Uplift, Tectonics, and Erosion of Eastern Tibet from Large-Scale Drainage Patterns," *Tectonics* 23, no.1 (2004): 1–20, https://doi.org/10.1029/2002TC001402; Peng Zhang et al., "Palaeodrainage Evolution of the Large Rivers of East Asia, and Himalayan-Tibet Tectonics," *Earth-Science Review* 92 (2019): 601–30. It is no less intriguing that in Chinese, the Red River is known as Yüan Chiang, meaning "Original River." While many rivers in Asia bear distinctive names, the originary significance of the Red River in Chinese cultural vocabulary sets it apart. The qualification of "original," as attached to the Red, roughly corresponds to some paleogeolgoical findings about her as a primitive source of other neighboring rivers.

5. Quoted in J. George Scott, ed., *Gazetteer of Upper Burma and the Shan States*, part. 1, vol. 1 (Government of British Burma, 1900), 11; Burrard and Hayden call this "geological convergence." See S. G. Burrard and H. H. Hayden, *A Sketch of the Geology*

and Geography of the Himalayan Mountains and Tibet, part III: *The Rivers of the Himalaya and Tibet* (Superintendent Government Press, 1907), 127–29.

6. Wang Gungwu, "A Two-Ocean Mediterranean," in *Anthony Reid and the Study of the Southeast Asian Past*, ed. Geof Wade and Tana Li (ISEAS-Yusof Ishak Institute, 2012), 47–68. Major works using Braudelian lens include K. N. Chaudhuri, *Trade and Civilizations in the Indian Ocean: An Economic History* (Cambridge University Press, 1985) and Anthony Reid, *Southeast Asia in the Age of Commerce*, vol. 1 (Yale University Press, 1988) and *Southeast Asia in the Age of Commerce*, vol. 2 (Yale University Press, 1993); Rila Mukherjee, ed., *Pelagic Passageways: The Northern Bay of Bengal Before Colonialism* (Primus Books, 2011). For studies on Indian Ocean space from intellectual and migration perspective, see Sugata Bose, *A Hundred Horizons: The Indian Ocean in the Age of Global Empire* (Harvard University Press, 2009); Sunil S. Amrith, *Crossing the Bay of Bengal: The Furies of Nature and the Fortunes of Migrants* (Harvard University Press, 2013).

7. Willem van Schendel, "Geographies of Knowing, Geographies of Ignorance: Jumping Scale in Southeast Asia," *Environment and Planning D: Society and Space* 20, no. 6 (2002): 654.

8. Fernand Braudel, *The Mediterranean and the Mediterranean World in the Age of Philip II*, vol. 1 (University of California Press, 2023), 40–41.

9. "Zomia" was first coined and explained by Willem van Schendel in "Geographies of Knowing, Geographies of Ignorance," 647–68. For a broader Braudelian application of the term, see James C. Scott, *The Art of Not Being Governed* (Yale University Press, 2010); Jean Michaud, ed., *Turbulent Times and Enduring Peoples: Mountain Minorities in the South East Asian Massif* (Routledge, 2000); Jean Michaud, "The Montagnards and the State in Northern Vietnam from 1802–1975: A Historical Overview," *Ethnohistory* 47, no. 2 (2000). On overlaps between Zomia and Southeast Asia massif, see Jean Michaud, "Zomia and Beyond," in *Routledge Handbook of Asian Borderlands*, ed. A. Horstmann, M. Saxer, and A. Rippa (Routledge, 2018).

10. The ten rivers are the Rhône, Po, Drin-Bojana, Nile, Neretva, Ebro, Tiber, Adige, Seyhan, and Ceyhan. See GRID-Arendal, "River Discharge of Freshwater into the Mediterranean," 2013, https://www.grida.no/resources/5897, accessed February 6, 2025; UN Global Compact, "Interactive Database of the World's River Basins", https://ceowatermandate.org/disclosure/resources/river-basins/, accessed February 6, 2025.

11. For a major study of the monsoonal lifeworld and the predicaments facing its rivers, see Sunil Amrith, *Unruly Waters: How Rains, Rivers, Coasts, and Seas Have Shaped Asia's History* (Basic Books, 2018).

12. Willem van Schendel, "Southeast Asia: An Idea Whose Idea Is Past?" *Bijdragen tot de Taal-, Land- en Volkenkunde* 168, no. 4 (2012): 497–503.

13. Prasenjit Duara, *The Crisis of Global Modernity* (Cambridge University Press, 2014).

14. Eric Tagliocozzo, Helen F. Siu, and Peter C. Perdue, eds., *Asia Inside Out: Connected Places* (Harvard University Press, 2015); Eric Tagliacozzo and Wen-Chin Chang, eds., *Chinese Circulations: Capital, Commodities, and Networks in Southeast Asia* (Duke University Press, 2011).

15. Karl Wittfogel, *Oriental Despotism: A Comparative Study of Total Power* (Yale University Press, 1957). For the specific case of intensive commercial and scientific uses of the Ganga and Yellow rivers, see Cunha, *The Invention of Rivers*; Elizabeth Whitcombe, *Agrarian Conditions in Northern India* (University of California Press, 1972); Zhang, *The River, the Plain, the State*. For an early assessment of ecological changes in the mid-Ganga basin in colonial India, see Michael Mann, *British Rule on Indian Soil: North India in the First Half of the Nineteenth Century* (Manohar, 1999).

16. Donald Worster, *Rivers of Empire: Water, Aridity, and the Growth of the American West* (Oxford University Press, 1992); Zhang, *The River, the Plain, the State*; Maya K. Peterson, *Pipe Dreams: Water and Empire in Central Asia's Aral Sea Basin* (Cambridge University Press, 2019); Jennifer L. Derr, *The Lived Nile: Environment, Disease, and Material Colonial Economy in Egypt* (Stanford University Press, 2019).

17. Janice Stargardt, "Hydraulic Works and South East Asian Polities," in *Southeast Asia in the 9th to 14th Centuries*, ed. D. Marr and A. Milner (ISEAS-Yusof Ishak Institute, 1986).

18. John Robison, *A System of Mechanical Philosophy*, vol. 2 (Edinburgh, 1822), 369, 372.

19. F. Kingdon Ward, *In Farthest Burma* (Seeley, Service and Co., 1921), 289.

20. Sean W. Fleming, *Where the River Flows: Scientific Reflections on Earth's Waterways* (Princeton University Press, 2017), 38.

21. For a note on "multivocal river," see Nianshen Song, *Making Borders in Modern East Asia: The Tumen River Demarcation 1881–1919* (Cambridge University Press, 2018). For the "audibility" of river and river-borne landscape, see Emerson's poem "The River." A long poem titled "Nodi" (River) by Tagore imagines the river to be calling everyone with "raised arms." Toni Morrison sees the flooding of the Mississippi not as a flood but a "remembering of where it used to be," because "All water has a perfect memory and is forever trying to get back where it was." See Toni Morrison, "The Site of Memory," in *Inventing the Truth: The Art and Craft of Memoir*, ed. William Zinsser (Houghton Mifflin, 1995), 96.

22. Stephen Rice, André Roy, and Bruce Rhoads, eds., *River Confluences, Tributaries and the Fluvial Network* (Wiley, 2008); Andrea Rinaldo, Marino Gatto, and Ignacio Rodriguez-Iturbe, *River Networks as Ecological Corridors: Species, Populations, Pathogens* (Cambridge University Press, 2020).

23. Bruno Latour, "On Actor-Network theory: A Few Clarifications," *Soziale Welt* 47, no. 4 (1996): 369–81.

24. K. V. Zvelebil, *The Lord of the Meeting Rivers. Devotional Poems of Basavanna* (Motilal Banarsidass; UNESCO, 1984), xx.

25. Burrand and Hayden, *A Sketch of the Geology and Geography of the Himalayan Mountains and Tibet*, 127–29. Even experienced travelers like T. T. Cooper confused the Yarlung Tsangpo with the Irrawaddy rather than the Brahmaputra, as late as 1869, despite James Rennell's identification earlier. See TBL, IOR/L/PS/6/564, part 152: Notes by T. T. Cooper on water communication between Hankow Chung-king and Chen-tu, the capital of Sz-chuen, 1. See also Thomas Simpson, "Find the River: Discovering the

Tsangpo-Brahmaputra in the Age of Empire," *Modern Asian Studies* 58, no. 1 (2023): 127–62.

26. Braudel, *The Mediterranean*, 134, 220.

27. Christof Mauch and Thomas G. Zeller, eds., *Rivers in History: Perspectives in Europe and North America* (University of Pittsburg Press, 2008), 7.

28. Toni Huber, "Micro-Migrations of Hill Peoples in Northern Arunachal Pradesh: Rethinking Methodologies and Claims of Origins in Tibet," in *Origins and Migrations in the Extended Eastern Himalayas*, ed. Toni Huber and Stuart Blackburn (Brill, 2012), 83.

29. Dinabandhu Mitra, *Kamale Kamini*, [কমলে কামিনী], (Nūtana Saṃskṛta Yantra, 1873), 362–63.

30. A. P. Phayre, "On the History of the Burmah Race," *Transactions of the Ethnological Society of London* 5 (1867): 31–32; Martin Gaenszle, "Where the Waters Dry Up—The Place of Origin in Rai Myth and Ritual," in *Origins and Migrations in the Extended Eastern Himalayas*, ed. Toni Huber and Stuart Blackburn (Brill, 2012), 43.

31. Jan Wissenman Christie, "Water and Rice in Early Java and Bali," in *A World of Water: Rain, Rivers and Seas in Southeast Asian Histories*, ed. Peter Boomgaard (KITLV, 2007), 235.

32. Pon Nya Mon, "Ethnic Identity and Political Autonomy of the Mon," in *The Mon Over Two Millennia: Monuments, Manuscripts, Movements*, ed. Patrick McCormick, Mathias Jenny, and Chris Baker (Institute of Asian Studies, 2011), 179–80; Nai Pan Hla, *The Significant Role of the Mon Language and Culture in Southeast Asia*, part 1 (Institute for the Study of Languages and Cultures of Asia and Africa, 1992), 7.

33. R. S. Lyngdoh, *John Robert's Ka Histori Ka Thoh Ka Thar* (H. W. Sten, 1979), 12.

34. Phayre, "On the History of the Burmah Race," 35.

35. Bin Yang, *Between Winds and Clouds: The Making of Yunnan* (Columbia University Press, 2008); Radhika Seshan, *Narratives, Routes and Intersections in Pre-Modern Asia* (Routledge, 2016), 44–50, 57.

36. Iftekhar Iqbal, "Reclaiming the Crossroads Between India and China: A View from the River," *Economic and Political Weekly* 49, no. 51 (2014): 20–23.

37. See Corey Ross's excellent recent book, *Liquid Empire* (Princeton University Press, 2024), 11.

38. J. R. McNeill, *The Mountains of the Mediterranean World: An Environmental History* (Cambridge University Press, 1992), 8.

39. F. Kingdon Ward, "The Valleys of the Kham," *The Geographical Journal* 56, no. 3 (1920): 192.

40. Anna Tsing, *Friction: An Ethnography of Global Connection* (Princeton University Press, 2004).

41. For debates on competing human encounters with the commons, see Garrett Hardin, "Tragedy of the Commons," *Science* 162, no. 3859 (1968): 1243–48; Elinor Ostrom, *Governing the Commons: The Evolution of Institutions for Collective Action* (Cambridge University Press, 1990).

42. Helmut R. Külz, "Further Water Disputes Between India and Pakistan." *The International and Comparative Law Quarterly* 18, no. 3 (July 1969): 725–26; For a discussion on the decommonization of South Asian rivers, see Iftekhar Iqbal, "Locating the Riparian Commons in Eastern South Asia: A Translocal Perspective", in *Urban Development and Environmental History in Modern South Asia*, ed. Ian Talbot and Amit Ranjan (Routledge, 2023).

43. David A. Pietz, *The Yellow River: The Problem of Water in Modern China* (Harvard University Press, 2015).

44. David Bollier and Silke Helfrich, eds., *Patterns of Commoning* (The Commons Strategies Group, 2015); Ash Amin and Philip Howell, "Thinking the Commons," in *Releasing the Commons: Rethinking the Futures of the Commons*, ed. Ash Amin and Philip Howell (Routledge. 2016); Prateep Kumar Nayak, ed., *Making Commons Dynamic: Understanding Change Through Commonisation and Decommonisation* (Routledge, 2021); Dan Smyer Yü, "Symbiotic Indigeneity and Commoning in the Anthropogenic Himalayas," in *Environmental Humanities in the New Himalayas: Symbiotic Indigeneity, Commoning and Sustainability*, ed. Dan Smyer Yü and Erik de Maaker (Routledge, 2021).

45. Michelle Ann Miller, Carl Middleton, Jonathan Rigg, and David Taylor, "Hybrid Governance of Transboundary Commons: Insights from Southeast Asia," *Annals of the American Association of Geographers* 110, no. 1 (2020): 297–313.

46. Yü, "Symbiotic Indigeneity and Commoning," 249.

47. As Jon Wilson shows in the case of India, rather than being fully in control, the empire's actions were fraught with anxiety, uncertainty, and resultant violence. See his *India Conquered: Britain's Raj and the Chaos of Empire* (Simon & Schuster, 2016).

48. Ruth Gamble, Gillian G. Tan, Hongzhang Xu, Sara Beavis, Petra Maurer, Jamie Pittock, et al., *Rivers of the Asian Highlands: From Deep Time to the Climate Crisis* (Routledge, 2025).

Chapter 1

1. The Brahmaputra is the highest river in the world. In Sanskrit, Brahmaputra means "Son of Brahma," the creator of the universe. The Brahmaputra's Tibetan segment is known as Yarlung Tsangpo, which, in its different etymological iterations, refers to its heavenly origin and also a great river of Yarlung region. In India, it is referred to as Sri Lohit, Lohit, or Luit in Assamese, and Siang in the Abor region. It is also known as Dihang. Its lower segment in Bangladesh is known as Jamuna.

2. James Rennell, "An Account of the Ganges and the Burrampooter Rivers," *Philosophical Transactions of the Royal Society of London* 71 (1781): 87–114. For precolonial trade relations between China, Assam, and Bengal via the Brahmaputra, see Bin Yang, *Between Winds and Clouds: The Making of Yunnan* (Columbia University Press, 2008).

3. For recent historical sketches of Bangladesh centering on the delta, see Willem van Schendel, *A History of Bangladesh* (Cambridge University Press, 2020) Iftekhar Iqbal, *The Bengal Delta: Ecology, State and Social Change 1840–1943* (Palgrave, 2010).

4. Jayeeta Sharma, *Empire's Garden: Assam and the Making of India* (Duke University Press, 2011). For other recent literature on Assam and India's northeast in general, see Sanjib Baruah, *In the Name of the Nation* (Stanford University Press, 2020); Sanghamitra Misra, *Spaces, Borders, Histories: Identity Construction in Colonial Goaalpara* (University of London, 2004). Arupjyoti Saikia's major work on the Brahmaputra focuses on a longer history of the midstream of the river and its agency. See Arupjyoti Saikia, *The Unquiet River: A Biography of the Brahmaputra* (Oxford University Press, 2019).

5. Berenice Guyot-Réchard, *Shadow States: India, China and the Himalayas, 1910–1962* (Cambridge University Press, 2017); Willem van Schendel, *Bengal Borderland: Beyond Stage and Nation in South Asia* (Anthem, 2004); Yasmin Saikia, *Fragmented Memories: Struggling to Be Tai-Ahom in India* (Duke University Press, 2004).

6. Iftekhar Iqbal, "Reclaiming the Crossroads Between India and China: A View from the River," *Economic and Political Weekly* 49, no. 51 (2014): 20–23; Gunnel Cederlöf, *Founding an Empire on India's North-Eastern Frontiers, 1790–1840: Climate Commerce, Polity* (Oxford University Press, 2013); David Ludden, "India's Spatial History in the Brahmaputra-Meghna River Basin", in *Landscape, Culture, and Belonging: Writing the History of Northeast India*, ed. Neeladri Bhattacharya and Joy L. K. Pachuau (Cambridge University Press, 2019); Dilip Gagoi, *Making of India's Northeast: Geopolitics of Borderland and Transnational Interactions* (Routledge, 2019).

7. Montgomery Martin, *The History, Antiquities, Topography and Statistics*, vol. 3 (W. H. Allen, 1838), 643; Peal lists at least 95 rivers with their names across upper, lower, and central Assam. See S. L. Peal, "A Peculiarity of the River Names in Assam and Some of the Adjoining Countries," *Journal of the Asiatic Society of Bengal* 48, no. 4 (1879): 261–70.

8. In the Patkai Hill ranges between Assam and Burma that formed the Brahmaputra-Chindwin watershed, a colonial official noted, "A peculiar feature of this country is the number of streams and rivers that flow at right angles to the course that they must eventually take to reach the Brahmaputra." See E. T. D. Lambert, "From the Brahmaputra to the Chindwin," *The Geographical Journal* 89, no. 4 (1937): 313.

9. TBL, IOR/M/4/2404: Report of Captain W. E. Cross, on the Lohit Valley Reconnaissance Through the Sadiya Frontier Tract, Assam, India, December 15, 1941–February 6, 1942.

10. F. Kingdon Ward, "The Seinghku and Delei Valleys, North-East Frontier of India," *The Geographical Journal* 75, no. 5 (May 1930): 423–24.

11. W. W Hunter, *The Imperial Gazetteer of India*, vol. 8 (Trübner and Company, 1886), 389.

12. Quoted in Iqbal, *The Bengal Delta*, 45; Sirajganj was described in 1906 as the "Chief river mart" on the Jamuna. See Government of India, *Imperial Gazetteer: Eastern Bengal and Assam. Mountains, Lakes, Islands, Rivers, Canals and Historic Areas* (The Bengal Secretariat Press, 1906), 4.

13. G. W. Protherto, ed., *Tibet* (H. M. Stationery Office, 1920), 54–55. For a more recent cartographic understanding of the Tibetan fluvial connections with the

Tsangpo and other smaller river systems across Tibetan-Indian borders, see Diana Lange, *An Atlas of the Himalayas by a 19th Century Tibetan Lama: A Journey of Discovery* (Brill, 2020),156–244.

14. T. T. Cooper to the President of the Bengal Chamber of Commerce, in "Communication Between Assam and China," *Journal of the Society of Arts* 18, no. 937 (November 4, 1870): 933.

15. The Lohit river took the name of Tellu after Brahma Kund, and between Minzong and Walong it took the name of Krawnaon. Beyond Rima it was joined by two tributaries named Zayul Chu (Chayu) and Rongto Chu. See TBL, IOR/M/4/2404: Report of Captain W. E. Cross on the Lohit Valley Reconnaissance.

16. Protherto, *Tibet*, 55. On the trade connections between Assam and Sichuan as seen through the annual fair in Sadiya, see House of Commons, *Parliamentary Papers: Accounts and Papers*, vol. 49 (1874), 118.

17. Ronald Kaulback, *Salween* (Hodder and Stoughton, 1938), 181.

18. Prince Henri d'Orléans, *Tonking to India* (Methuen & Co.,1898), 305, 315, 324.

19. Protherto, *Tibet*, 68–69.

20. C. R. Markham, "Travels in Great Tibet, and Trade Between Tibet and Bengal," *The Journal of the Royal Geographical Society of London* 45 (1875): 302.

21. Markham, "Travels in Great Tibet," 313–14.

22. Somerset Playne, *Bengal and Assam, Behar and Orissa: Their History, People, Commerce and Industrial Resources* (Foreign and Colonial Compiling and Publishing Co., 1917), 207.

23. Arthur Cotton, "On a Communication Between India and China by the Line of the Burhampooter and Yangtsze," *Journal of the Royal Geographical Society of London* 37 (1867): 231–39.

24. TBL, IOR/L/PS/6/56, Part 152: T. T. Cooper to the President of the Bengal Chamber of Commerce, July 29, 1869.

25. Iqbal, "Reclaiming the Crossroads."

26. TBL, IOR/L/PS/6/564, Part 152: T. T. Cooper to Under Secretary of the Government of India, Foreign Dept, Memorandum on a trader route between India and China via Zyyu and Bathang, with its advantage, 1–3.

27. J. McSwiney, *Census of India*, 1911, vol. 3, Assam, Pt. 1, Report (Assam Secretariat Printing Office, 1912), 159, 175.

28. Saikia, *The Unquiet River*, 137.

29. Iftekhar Iqbal, "The Space Between Nation and Empire: The Making and Unmaking of Eastern Bengal and Assam, 1905–1911," *The Journal of Asian Studies* 74, no. 1 (2015).

30. Noël Williamson, "The Lohit-Brahmaputra Between Assam and South-Eastern Tibet, November, 1907, to January, 1908," *The Geographical Journal* 34, no. 4 (1909): 383.

31. TBL, IOR/L/PS/6/564 PARTS 152: "Slip of Meeting of the Royal Geographical Society of May 3, 1870," 9.

32. TBL, IOR/L/PS/6/564: part 152: Cooper's talk at Royal Geographic Society, July 13, 1870: "Travels in Western China and Eastern Thibet."

33. TBL, IOR/M/4/2404: Copy of a note by E. D. T. Lamberts, Central Intelligence Officer, Assam. September 21, 1945.

34. TBL, IOR/L/PS/6/564: part 152, E. H. Higgs to T. T. Cooper, July 29, 1869.

35. TBL, IOR/M/4/2404: Report of Captain W. E. Cross, on the Lohit Valley Reconnaissance Through the Sadiya Frontier Tract, Assam, India, December 15, 1941–February 6, 1942.

36. For the concept of the empire as a process, see David Ludden, "The Process of Empire: Frontiers and Borderlands," in *Tributary Empires in Global History*, ed. Peter Fibiger Bang and C.A. Bayly (Springer, 2020).

37. Powell Millington, *On the Track of the Abor* (Smith, Elder & Co., 1912), 203.

38. Guyot-Réchard, *Shadow States*, 31–92; For an assessment of Indo-China frontier relations as an evolving process of geopolitics around the western Himalaya-Tibet region, see Kyle J. Gardner, *The Frontier Complex: Geopolitics and the Making of the India-China Border* (Cambridge University Press, 2021). These books offer a useful account of India and China's border outreaches that shaped the postcolonial borders.

39. Anonymous, "The North-East Frontier of Bengal," *Saturday Review of Politics, Literature, Science and Art* 58 (October 25, 1884): 536–37.

40. Protherto, *Tibet*.

41. Bodhisattva Kar, "Nomadic Capital and Speculative Tribes: A Culture of Contracts in the Northeastern Frontier of British India," *The Indian Economic and Social History Review* 53, no. 1 (2016): 41–67.

42. Millington, *On the Track of the Abor*, 212.

43. TBL, Mss Eur F157/208: Diary of Tsangpo Expedition (April 25 to November 14, 1913), 25–26.

44. T. T. Cooper, *The Mishmee Hills: An Account of a Journey Made in an Attempt to Penetrate Thibet from Assam to Open New Routes for Commerce* (Henry S. King & Co. 1873), 128.

45. TBL, IOR/L/PS/6/564, part 152: E. H. Higgs to T. T. Cooper, July 29, 1869.

46. TBL, IOR/L/PS/6/564, part 152: Translation of Memorial addressed by Thibetan authorities to Pekin.

47. TBL, MSS Eur. F157/304C: Journey of a Diary from Peking to Assam, by Captain F. M. Bailey (1911), 61.

48. TBL, IOR/L/PS/6/564, part 152: T. T. Cooper to President of the Bengal Chamber of Commerce, July 29, 1869; TBL, IOR/L/PS/564, part 152: Translation of Memorial addressed by Thibetan authorities to Peking, translated through Thibetan and Manchee into Chinese, 2: "Hsitsang (Thibet) is the native country of Buddhism."

49. F. Kingdon Ward and Malcolm Smith, "The Himalaya East of the Tsangpo," *The Geographical Journal* 84, no. 5 (November 1934): 380.

50. Cooper, *The Mishmee Hills*, 129.

51. Ward and Smith, "The Himalaya East of the Tsangpo," 374.

52. L. W. Shakespear, *History of Upper Assam, Upper Burmah and North-Eastern Frontier* (Macmillan and Co., 1914), 113.

53. TBL, IOR/L/PS/11/15: Diary of the Assistant Political Officer, Abor Expeditionary Force, from February 25 to March 19, 1912, 1–2.

54. TBL, IOR/L/PS/11/57: G. A. Nevill, Political Officer, Dibong Survey to Chief Secretary to Chief Commissioner of Assam, May 15, 1913.

55. TBL, IOR/L/PS/12/3111: W. A. Cosgrave, Assam Secretariat to E. B. Howell, Foreign Secretary to the Govt. of India, September 10, 1930; and E. B. Howell to W. A. Cosgrave, July 8, 1930.

56. Shakespear, *History of Upper Assam*, 121.

57. Government of Great Britain, *Parliamentary Papers, Great Britain, vol. 49, House of Commons: Accounts and Papers*, 1873, 161; Cooper notes: "If the English could open a road through the Mishmee country they would find at Dzayul the old Mandarin Road coming from Yun-nan," which was the closest to the upper Brahmaputra of three routes running from China to India: from Dali by "Kiang," entering Tibet near "Atentze, crossing the districts of Tsarange of Dzayul at the foot of the Himalaya near the Mishmees and south of Pomi." See T. T. Cooper, "Notes on Tibet, by a French Missionary," in "Letter from Mr. T. T. Cooper, on the Course of the Tsan-Po and Irrawaddy and on Tibet," *Proceedings of the Royal Geographical Society of London* 13, no. 5 (1868–1869): 394.

58. J. P. Mills, "The Mishmis of the Lohit Valley, Assam," *The Journal of the Royal Anthropological Institute of Great Britain and Ireland* 82, no. 1 (1952): 1; see also Figure 1.1.

59. Ambika Aiyadurai and Claire Seungeun Lee, "Living on the Sino-Indian Border: The Story of the Mishmis in Arunachal Pradesh, Northeast India," *Asian Ethnology* 76, no. 2 (2017): 367–95; J. P. Mills mentioned about 120 different Mishmi clans. See J. P. Mills, "The Mishmis of the Lohit Valley, Assam."

60. TBL, IOR/M/4/2404: Report of Captain W. E. Cross, on the Lohit Valley Reconnaissance, 21.

61. TBL, Mss Eur F157/208: Diary of Tsangpo Expedition (April 25 to November 14, 1913), 5–6.

62. Cooper, *The Mishmee Hills*, 184.

63. Cooper, *The Mishmee Hills*, 183. *Teta* was a decoction of the roots of the *Coptis teeta* plant, grown around 7,000–9,000 feet from the sea level, for use as tonic and febrifuge. The roots were also brought to Sadiya where these were "eagerly bought" by Assamese and Bengali merchants. See Cooper, *The Mishmee Hills*, 214; See also Mills, "The Mishmis of the Lohit Valley, Assam," 5.

64. Cooper, *The Mishmi Hills*, 218.

65. Cooper, *The Mishmi Hills*, 232, 244, 247.

66. Report of Captain W. E. Cross, on the Lohit Valley Reconnaissance, 19.

67. Report of Captain W. E. Cross, on the Lohit Valley Reconnaissance.

68. Report of Captain W. E. Cross, on the Lohit Valley Reconnaissance, 3, 7, 12, 18.

69. Henry Cottam, "Overland Route to China via Assam, Tenga Pani River, Khamti, and Singphoo Country," *Proceedings of the Royal Geographical Society of London* 21, no. 6 (1876–1877): 592.

70. Cooper, *The Mishmee Hills*, 236–37.
71. TBL, IOR/L/PS/11/57: G. A. Nevill: Dibong Survey to Chief Secretary to Chief Commissioner to Assam.
72. Kaulback, *Salween*, 59.
73. TBL, IOR: L/PS/11/15: Diary of the Assistant Political Officer, Abor Expeditionary Force, from February 25 to March 19, 1912, 5.
74. Ward and Smith, "The Himalaya East of the Tsangpo," 386.
75. TBL, MSS Eur. F157/208: Diary of Tsangpo Expedition (April 25 to November 14, 1913) 25–26.
76. d'Orléans, *Tonkin to India*, 294, 303.
77. d'Orléans, *Tonkin to India*, 343.
78. Quoted in Hunter, *The Imperial Gazetteer of India*, 547; Iftekhar Iqbal, "The Bengali Muslim: Language and Space-Making at the Ocean's Margins," in *Oceanic Islam*, ed. Sugata Bose and Ayesha Jalal (Bloomsbury, 2020), 183.
79. Swarupa Gupta, *Cultural Constellations, Place-Making and Ethnicity in Eastern India, c. 1850–1927* (Brill, 2017), 70, 151–52.
80. Lakshmīnātha Bejabaruwā, *jivan smoron* (Assam Book Trust, 1992), 3.
81. Iqbal, "The Bengali Muslim"; Gyanendra Mohon Das, *Banger Bahire Bangali* [The Bengalis Beyond Bengal] (Indian Publishing House, 1931), 384.
82. Baruah, *In the Name of the Nation*.
83. Cooper, *The Mishmi Hills*, 3–4, 134–35.
84. Arupjyoti Saikia, "Imperialism, Geology and Petroleum: History of Oil in Colonial Assam," *Economic and Political Weekly* 46, no. 12 (2011); Sharma, *The Empire's Garden*.
85. Ambika Aiyadurai, "The Meyor: A Least Studied Frontier Tribe of Arunachal Pradesh, Northeast India," *The Eastern Anthropologist* 64, no. 4 (2011): 464–65.
86. Heiko Schrader, "A Himalayan Trading Community in Southeast Asia," in *The Moral Economy of Trade: Ethnicity and Developing Markets*, ed. Hans-Dieter Evers and Heiko Schrader (Routledge, 1994).
87. Cooper, *The Mishmi Hills*, 150, 230.
88. Cottam, "Overland Route to China," 590; Cooper, *The Mishmee Hills*, 37.
89. Kaulback, *Salween*, 121–23.
90. Lambert, "From the Brahmaputra to the Chindwin," 310.
91. Lambert, "From the Brahmaputra to the Chindwin," 318, 323.
92. Jelle Wouters, "Keeping the Hill Tribes at Bay: A Critique from India's Northeast of James C. Scott's Paradigm of State Evasion," *European Bulletin of Himalayan Research* 39 (2011); Bengt G. Karlsson, "Evading the State: Ethnicity in Northeast India Through the Lens of James Scott," *Asian Ethnology* 72, no. 2 (2013): 321–33.

Chapter 2

1. In Sanskrit, Irrawaddy or Ayeyarwady means "Elephant-like." The term is also translated as "abounding in riches." A tributary of the river inside Yunnan is known as Dulongjiang, which literally means "Solitary Dragon River." However, the name is primarily derived from the Dulong people (Dúlóngzú), an ethnic minority in the region.

2. Henry M. Cadell, "A Sail Down the Irrawaddy," *Scottish Geographical Magazine* 15, no. 5 (1901): 240.

3. Thant Myint-U, *Where China Meets India: Burma and the New Crossroads of Asia* (Faber and Faber, 2012), 292.

4. Henry Yule, "On the Geography of Burma and Its Tributary States, in Illustration of a New Map of Those Region," *The Journal of the Royal Geographical Society of London* 27 (1857): 65n.

5. E. B. Sladen, "Expedition from Burma, via the Irrawaddy and Bhamo, to South-Western China," *The Journal of the Royal Geographical Society of London* 41 (1871): 276, 279.

6. TBL, IOR/L/PS/6/564, part 152: J. Anderson, "Irrawaddy and Its Sources," paper read at the Royal Geographical Society, June 13, 1870; F. Kingdon Ward, *In Farthest Burma* (Seeley, Services & Company, 1921), 240; d'Orléans identified three routes from Khampti to Assam: southward, lengthy and more frequented; northward in the direction of Zayul through the Mishimi countries toward Brahmaputra; westward, mountainous but shortest. See Prince Henry d'Orléans, "From Yun-nan to British India," *The Geographical Journal* 7, no. 3 (March 1896): 306.

7. B. E. A. Pritchard, "A Journey from Myitkyina to Sadiya viâ the N'mai Hka and Hkamti Long," *The Geographical Journal* 43, no. 5 (May 1914): 535.

8. Ward, *In Farthest Burma*, 19, 159n; *Rame* is a Kachin word for small river and *zup* stands for smaller stream. *Hka* denotes a river larger than a *rame*.

9. Karl Andree, *Geographic des welthandels: Mit geschichtlichen erläuterungen*, vol. 2 (Berlag von Julius Meier, 1872), 381.

10. Ward, *In Farthest Burma*, 159n.

11. Ward, *In Farthest Burma*, 24.

12. L. W. Shakespear, *History of Upper Assam, Upper Burmah and North-Eastern Frontier* (Macmillan and Co., 1914), 56.

13. Michael Adas, *Burma Delta: Economic Development and Social Change on an Asian Rice Frontier 1852–1941* (University of Washington Press, 2011); Willem van Schendel, "Origin of Burma Rice Boom," *Journal of Contemporary Asia* 17, no. 4 (1987); Ian Brown, *A Colonial Economy in Crisis* (Routledge, 2005).

14. Aung Kyaw Htoo "Myanmar Petroleum Sector Future Pathways and Prospect," (Ministry of Energy, 2014), 4. https://web.archive.org/web/20150417055259/http://www.myanmar-oilgas.com/OilGas/media/Site_Images/Day-2-AM-1-EPD-Petroleum-Sector-Prospects.pdf. Accessed May 9, 2025.

15. John M'Cosh, "On the Various Lines of Overland Communication Between India and China," *Proceedings of the Royal Geographical Society of London* 5, no. 2 (1860–1861): 49–50.

16. Yule, "On the Geography of Burma," 69.

17. John L. Christian, "Trans-Burma Trade Routes to China," *Pacific Affairs* 13, no. 2 (1940): 179; see also TBL, IOR/L/PS/6/555, no. 88: Chief Commissioner, British Burma to Secretary to Government of India, Foreign Dept, February 1, 1868.

18. TBL, IOR/L/PS/18/B117: Memorandum on Indian Trade with Western China, and on the Resources, Trade, and Trade Route of Yunnan and Adjacent Provinces, 1899, 5.

19. Shakespear, *History of Upper Assam*, 183.

20. TBL, IOR/L/PS/18: Schemes of Railway Extension in China, 3.

21. TBL, Mss Eur E375/2 (a&b): The Irrawaddy Flotilla & Burmese Steam Navigation Company, Limited. Glasgow, 1872, 5; The Panama Canal was already under discussion, heightening British concerns about American commercial competition and reinforcing their urgency to establish exclusive dominance over the Irrawaddy corridor. See Christian, "Trans-Burma Trade Routes to China", 180.

22. Alister McCrae and Alan Prentice, *Irrawaddy Flotilla* (James Paton Limited, 1978), 66.

23. The Irrawaddy Flotilla & Burmese Steam Navigation Company, 6.

24. The Irrawaddy Flotilla & Burmese Steam Navigation Company, 4.

25. The Irrawaddy Flotilla and the Burmese Steam Navigation Company, 6, 11–12, 14.

26. Than Tun, ed., *Wekmasuk Wundauk U Latt's Diary, 1888–1889*, vol. 2, trans. Myo Oo (Busan University of Foreign Studies Press, 2014), 225, 229, 217.

27. TBL, IOR/v/27/732/25: Report on the Waterways in the Irrawaddy Delta (Office of the Superintendent (Government Printing Press, 1915), 17–18.

28. Report on the Waterways in the Irrawaddy Delta, 3.

29. Iftekhar Iqbal, "The Boat Denial Policy and the Great Bengal Famine," *Journal of the Asiatic Society of Bangladesh* 56, 1–2 (2011): 271–82.

30. For a translated version of the Chinese letter that characterized the English in such a way, see Yule, "On the Geography of Burma," 56–57.

31. Yi Li, "Transformation of the Yunnanese Community Along the Sino-Burma Border During the Nineteenth and Early Twentieth Centuries," in *Imperial China and Its Southern Neighbours*, ed. Victor H. Mair and Liam Kelley (ISEAS-Yusof Ishak Institute, 2018) 295.

32. R. Boileau Pemberton, "Abstract of the Journal of a Route Travelled by Captain S. F. Hannay, of the 40th Regiment, Native Infantry, in 1835–36, from the Capital of Ava to the Amber Mines of the Hukong Valley on the South-east Frontier of Assam," *SOAS Bulletin of Burma Research*, 3, no. (Spring 2005): 200.

33. TBL, IOR/L/PS/18, B.65–81: Burma-China Frontier Negotiations, October 18, 1893, 8, 29.

34. TBL, IOR/L/PS/18, B.65–81: Burma-China Frontier Negotiations, December 20, 1892.

35. TBL, Mss Eur F278/95: Diary of J. C. Scott, Her Majesty's Commissioner, Burma-China Boundary Commission, to the period ending February 25, 1899.

36. TBL, IOR/L/PS/12/4622: J. G. Laithwaite, to Raibeart MacDougall, March 7, 1947. This reflects the imperial state's ambition, rooted in its plains-based power, to assert influence in its highland peripheries in the postcolonial era—an argument Eric Tagliacozzo suggests as a critique of highland exceptionalism. See Eric Tagliacozzo,

"Ambiguous Commodities, Unstable Frontiers: The Case of Burma, Siam, and Imperial Britain, 1800–1900," *Comparative Studies in Society and History* 46, no. 2 (2004).

37. For details on the powerplays of India and China in these border regions, see Berenice Guyot-Réchard, *Shadow States: India, China and the Himalayas, 1910–1962* (Cambridge University Press, 2017).

38. J. S. Furnivall, *Colonial Policy and Practice: A Comparative Study of Burma and Netherlands India* (Cambridge University Press, 1948). Appendix II: Distribution of Seaborne Trade of Burma, 1869–1937.

39. A. R. MacMahon, "Burmese Border Tribes and Trade Routes," *Blackwoods Edinburgh Magazine* 140, no. 851 (September 1886): 403, 406.

40. Edward B. Sladen and Horace Browne, *Mandalay to Momien: A Narrative of the Two Expeditions to Western China of 1868 and 1875* (Macmillan and Co, 1876), 8.

41. Sladen, "Expedition from Burma," 264.

42. Sladen and Browne, *Mandalay to Momien*, 327.

43. Edmund Leach, *Political System of Highland Burma* (Oxford University Press, 1968), 32.

44. TBL, IOR/V/27/732/27: Report on the Ferries on the Mali Hka between Nonghkai Village in the Putao District and the Mali Hka-N'Mai Hka confluence in the Myitkyina District (1916), 1–6.

45. James C. Scott, *The Art of Not Being Governed* (Yale University Press, 2010), 22, 151, 248; Leach, *Political System of Highland Burma*, 21, 88.

46. Sladen and Browne, *Mandalay to Momien*, 128.

47. Wen-Chin Chang, "From a Shiji Episode to the Forbidden Jade Trade During the Socialist Regime in Burma," in *Chinese Circulations: Capital, Commodities, and Networks in Southeast Asia*, ed. Eric Tagliacozzo and Wen-Chin Chang (Duke University Press, 2011), 464.

48. Sladen and Browne, *Mandalay to Momien*, 152–3.

49. V. C. Scott O'Connor, *The Silken East: A Record of Life and Travel in Burma*, vol. 1 (Hutchinson & Co. 1904), 189.

50. TBL, Mss Eur F278/95: "Diary of [J. S. Scott] Her Majesty's Commissioner, Burma-China Boundary Commission, to the period ending 25th February 1899." 1–2.

51. TBL, Mss Eur E254/5: H. Thirkell White to Earl of Minto, Viceroy and Governor-General of India, March 18, 1906.

52. Sladen, "Expedition from Burma," 280.

53. TBL, IOR/L/PS/18. B.65–81: Burma-China Frontier Negotiations, September 8, 1892, 9.

54. TBL, IOR/L/PS/12/4622: Notes on the Chinese Shan States, March 3, 1943.

55. TNA, 1941: Pack animals. PPMS2, confidential letters, 1882–1941: Commissioner, Custom House, Tengyueh to Inspector General of Customs, Peking, July 29, 1907.

56. Ney Elias, "A Visit to the Valley of the Shueli, in Western Yunnan" (February 1875), *Proceedings of the Royal Geographical Society of London* 20, no. 4 (1875–1876), 238.

57. Shakespear, *History of Upper Assam*, 150.

58. Singpho dialects were found to be spoken in as far west on the Brahmaputra River Valley as Dibrugarh. See Ola Hanson, *The Kachins: Their Customs and Traditions* (first published 1913; Cambridge University Press, 2012), 12.

59. A. R. MacMahon, "Burmese Border Tribes," 402; Shakespear, *History of Upper Assam*, 152; Ola Hanson, *The Kachins*, 12, 35.

60. U. Chit Hlaing, "Anthropological Communities of Interpretation for Burma: An Overview," *Journal of Southeast Asian Studies* 39, no. 2 (June 2008): 245.

61. Eric Tagliacozzo and Wen-Chin Chang, "The Arc of Historical Commercial Relations Between China and Southeast Asia," in *Chinese Circulations: Capital, Commodities, and Networks in Southeast Asia*, ed. Eric Tagliacozzo and Wen-Chin Chang (Duke University Press, 2011), 1–16.

62. During his trip O'Connor met fifty-three such passengers who were either Shans or Laos. See O'Connor, *The Silken East*, 79.

63. J. W. Palmer, *Up and Down the Irrawaddi or The Golden Dagon: Being Passages of Adventure in the Burman Empire* (Rudd & Carleton, 1859), 34.

64. On aspects of Rangoon cosmopolitanism, see Devleena Ghose, "Burma-Bengal Crossings: Intercolonial Connections in Pre-Independence India," *Asian Studies Review* 40, no. 2 (2016): 156–72; Sana Aiyar, "Revolutionaries, Maulvis, and Monks: Burma's Khilafat Moment," in *Oceanic Islam: Muslim Universalism and European Imperialism*, ed. Sugata Bose and Ayesha Jalal (Bloomsbury Publishing India, 2020); Iftekhar Iqbal, "The Bengali Muslim: Language and Space-Making at the Ocean's Margins," in *Oceanic Islam*, ed. Bose and Jalal; Sunil Amrith, *Crossing the Bay of Bengal: The Furies of Nature and the Fortunes of Migrants* (Harvard University Press, 2013).

65. Tun, *Wekmasuk Wundauk U Latt's Diary*, 147; See also Amitav Ghose's novel, *The Glass Palace* (Random House, 2000).

66. Yule, "On the Geography of Burma," 69.

67. George Ernest Morrison, *An Australian in China* (Methuen & Co., 1895), 284.

68. O'Connor, *The Silken East*, 190–93, 199–200.

69. O'Connor, *The Silken East*, 204.

70. Shakespear, *History of Upper Assam*, 193–94.

71. Ward, *In Farthest Burma*, 139.

72. Ward, *In Farthest Burma*, 61.

73. F. Kingdon Ward, "The Seinghku and Delei Valleys, North-East Frontier of India," *The Geographical Journal* 75, no. 5 (May 1930): 413–14.

74. Pritchard, "A Journey from Myitkyina to Sadiya," 526.

75. Ward, *In Farthest Burma*, 150.

76. Christian Daniels and Jianxiong Ma, "Introduction: The Agency of Local Elites in the Transformation of Western Yunnan During the Ming Dynasty," in *The Transformation of Yunnan in Ming China: From the Dali Kingdom to Imperial Province*, ed. Christian Daniels and Jianxiong Ma (Routledge, 2020), 14.

77. Yule. "On the Geography of Burma," 80n.

78. d'Orléans, "From Yun-nan to British India," 305.

79. d'Orléans, "From Yun-nan to British India," 306.

80. Pritchard, "A Journey from Myitkyina to Sadiya," 531. Pritchard narrates how Naingvaws nurtured sago palm tree, 70 feet tall and a 7-foot girth at their full maturity, to use only during famine situations (531–33).
81. Ward, *Farthest Burma*, 178.
82. Pritchard, "A Journey from Myitkyina to Sadiya," 531.
83. Pritchard, "A Journey from Myitkyina to Sadiya," 535.

Chapter 3

1. The Salween River has various etymological variations in different cultural spaces it spans. The English name Salween is derived from the Burmese Thanlwin or Mon Thanlan, both of which roughly means "vigorous waterway" or "vital waterway." In Tibetan, the river is known as Gyalmo Ngulchu, meaning "Queen of Wealth," while its Chinese name Nu Jiang translates to "Angry River".
2. TBL, MSS Eur F278/101: J. George Scott, Superintendent and Political Officer of the Southern Shan States, to the Chief Secretary to the Government of Burma, March 20, 1903.
3. Holt S. Hallett, *A Thousand Miles on an Elephant in the Shan States* (William Blackwood and Sons, 1890), 53–54.
4. Government of India, *Imperial Gazetteer of India*, vol. 21 (Clarendon, 1908), 423; Thongchai Winichakul, *Siam Mapped: A History of the Geo-Body of a Nation* (University of Hawaii Press, 1997), 68–69. See also Paiboon Hengsuwan, "Explosive Border: Dwelling, Fear and Violence on the Thai-Burmese Border Along the Salween River," *Asia-Pacific Viewpoint* 54, no. 1 (2013): 109–22.
5. Amnuayvit Thitibordin, "Control and Prosperity: The Teak Business in Siam 1880s–1932" (PhD diss., University of Hamburg, 2016); Carl Middleton and Vanessa Lamb, eds., *Knowing the Salween River: Resource Politics of a Contested Transboundary River* (Springer Open, 2019).
6. Sai Aung Tun, *History of the Shan State: From Its Origin to 1962* (Silkworm Books, 2009), 31.
7. J. Coryton and Mr. Margary, "Trade Routes Between British Burmah and Western China; with Extracts of Letters from Mr. Margary," *Proceedings of the Royal Geographical Society of London* 19, no. 4 (1874–1875): 272.
8. R. G. Woodthorpe, "The Country of the Shans," *The Geographical Journal* 7, no. 6 (June 1896). *Chaung* is "stream" in Burmese.
9. Peter Lund Simmonds and William Henry Giles Kingston, eds., *Colonial Magazine and East India Review* 3 (September–December 1844): 195–98; For more data on Moulmein trade, see P. E. Jamieson, *Burma Gazetteer: Amherst District*, vol. A (Government Printing and Stationery, 1913).
10. G. J. Younghusband, *Eighteen Hundred Miles on a Burmese Tat: Through Burmah, Siam and the Eastern Shan States* (Asian Educational Services, 1995; imprint of W. H. Allen and Co., 1888), 35, 43–44, 46, 51.
11. Hallett, *A Thousand Miles on an Elephant*; also quoted in Andrew Forbes and David Henley, *The Haw: Traders of the Golden Triangle* (Asia Film House, 1997).

12. R. H. F. Sprye, "Communication with the South-West Provinces of China from Rangoon in British Pegu," *Proceedings of the Royal Geographical Society of London, Session 1860–61*: 47.

13. Sprye, "Communication with the South-West Provinces," 46; For a note on Sprye's upper Salween route to reach China via Kiang Hung on the Mekong and a subsequent proposal for a trade route to Simao across the Karenni Hills and Kiang Hung, see John L. Christian, "Trans-Burma Trade Routes to China," *Pacific Affairs* 13, no. 2 (1940), 177–78.

14. TBL, IOR/V/27/560/79: John Coryton, *Letter to the Liverpool Chamber of Commerce on the Prospects of a Direct Trade Route to China Through Moulmein* (T. Whittam at the Advertiser Press, 1870), 38.

15. Younghusband, *Eighteen Hundred Miles*, 75.

16. TBL, IOR/H/675: The Burmese War 1823–1828;

17. TBL, IOR: L/PS/533, collection 1: A. P. Phayre, chief comm of British Burma to H. M. Durand, Sec to GoI, Foreign Dept, March 28, 1864. For a detailed note on the teak trade, see Amnuayvit Thitibordin, "Control and Prosperity: The Teak Business in Siam 1880–1932" (PhD diss., University of Hamburg, 2016).

18. Coryton, *Letter to the Liverpool Chamber of Commerce*, 13.

19. Anonymous, "India and China," *The Times*, January 17, 1853, issue 21327, 5.

20. Henry Duckworth, *New Commercial Route to China (Capt. Sprye's Proposition)* (George Philip and Son, 1861), 8–28.

21. TBL, IOR: L/PS/533, collection 1: Woongyee (Commissioner) of the Martaban and Tenasserim Provinces (under British jurisdiction) to the Tsanbwa of Mobyani, nd.

22. TBL, IOR: L/PS/533, collection 1: Journal of Lieutenant G. Colquhoun Scone, While Employed on the Salween Surveying Expedition (Military Orphan Press, 1865), 73.

23. Coryton and Margary, "Trade Routes Between British Burmah and Western China," 272.

24. Coryton, *Letter to the Liverpool Chamber of Commerce*, 3.

25. Coryton, *Letter to the Liverpool Chamber of Commerce*, 4–5.

26. TBL, IOR: L/PS/533, collection 1: A. P. Phayre, Chief Commissioner of British Burmah to H. M. Durand, Secretary to Government of India, Foreign Department, March 28, 1864.

27. TBL, IOR/L/PS/533, collection 1: J. T. Wheeler, Assistant Secretary to Government of India, Foreign Department to Chief Commissioner, April 15, 1864.

28. TBL, IOR/L/PS/18. B.65–81: Burma-China Frontier Negotiations, February 17, 1893.

29. Quoted in Thomas E. McGrath, "A Warlord Frontier: The Yunnan-Burma Border Dispute, 1910–1937," *Organization of American Historians (OAH) Proceedings* (2003): 12.

30. F. Kingdon Ward, *In Farthest Burma* (Seeley, Services & Company, 1921), 283.

31. J. Coryton, "Trade Routes Between British Burmah and Western China." *Journal of the Royal Geographical Society of London* 45 (1875): 229–30.

32. Khu Oo Reh, "Karenni People at a Glance," in *Citizenship in Myanmar: Ways of Being in and from Burma*, ed. Ashley South and Marie Lall (Cambridge University Press, 2017), 140.

33. Government of Burma, *Burma Gazetteer, Salween District, vol. A* (Government Printing and Stationery Press, 1957), 2.

34. TBL, IOR: L/PS/533, collection 1: Journal of Lieutenant G. Colquhoun Sconce while employed on Salween Surveying expedition.

35. A. R. MacMahon, "Burmese Border Tribes and Trade Routes," *Blackwoods Edinburgh Magazine* 140, no. 851 (September 1886): 402–3.

36. Thitibordin, "Control and Prosperity," 177.

37. Quoted in Coryton, *Letter to the Liverpool Chamber of Commerce*, 9.

38. Archibald Ross Colquhoun, *Ethnic History of the Shans* (Manas Publications, 1985), 23.

39. TBL, IOR/L/PS/6/564 Part 143: Memo of Captain C. E. Watson, Deputy Commissioner Shwegyeen District to Commissioner of the Tenasserim Division, Moulmein, Camp Kyouk-nyat, March 1, 1869, and March 4, 1869.

40. Yin-Tang Chang, "Anthropological Features of the Shans and Their Geographical Environment in South-West Yunnan," *Man* 44, no. 55 (May–June 1944): 61–62.

41. L. W. Shakespear, *History of Upper Assam, Upper Burmah and North-Eastern Frontier* (Macmillan and Co., 1914), 168.

42. MacMahon, "Burmese Border Tribes and Trade Routes," 394–98.

43. TBL, Mss Eur F278/71: *Report on the Administration of the Shan States 1889–90*, 25.

44. Coryton and Margary, "Trade Routes Between British Burmah and Western China," 266, 277.

45. R. G. Woodthorpe, "Explorations on the Chindwin River, Upper Burma," *Proceedings of the Royal Geographical Society and Monthly Record of Geography*, New Monthly Series 11, no. 4 (April 1889): 216.

46. Younghusband, *Eighteen Hundred Miles on A Burmese Tat*, 15.

47. Chang, "Anthropological Features of the Shans," 65

48. W. F. B. Laurie, "British and Upper Burma, and Western China: Their Concurrent Commercial Interests," *Journal of the Society of Arts* (June 11, 1880): 643–44.

49. TBL, IOR/L/PS/18/B117: Memorandum on Indian Trade with Western China, and on the Resources, Trade, and Trade Route of Yunnan and Adjacent Provinces, 1899, 8.

50. Hallett, *A Thousand Miles on an Elephant*, 171.

51. George Litton, *Report on a Journey in North and North-West Yunnan, Season 1902–1903* (Shanghai Mercury, 1903), 24.

52. Quoted in Wen-Chin Chang, "Circulations via Tangyan, a Town in the Northern Shan State of Burma," in *Asia Inside Out: Connected Places*, ed. Eric Tagliacozzo, Helen F. Siu, and Peter Perdue (Harvard University Press, 2015), 243.

53. Chang, "Anthropological Features of the Shans," 62; Ward, *Farthest Burma*, 43; For notes on the wide-ranging mobility of the Shans, see also Chang, "Circulations via Tangyan," 243.

54. TBL, Mss Eur F278/78: Report on an Expedition to Manglun and the Wild Wa Country by J. G. Scott, Superintendent of the Northern Shan States: E. S. Symes, Chief Secretary to the Chief Commissioner, to the Secretary of the Government of India, Foreign Department, Rangoon, August 3, 1893.

55. For example, the colonial administration entered into such collaboration with 23 chiefs controlling 25,000 people from 261 villages spanning a number of tributaries and branches of the Salween and in the middle of gold mine tracts. See Mss Eur F278/78: "Pacification of the West Mang Lun with Notes on the Wild Wa Country" by J. G. Scott. Lashio, June 13, 1893, 7–8.

56. Report on an Expedition to Manglun and the Wild Wa Country.

57. G. Colquhoun Scone, *Journal of Lieut. G. Colquhoun Scone, While Employed on the Salween Surveying Expedition* (Military Orphan Press, 1865), 73.

58. For data on population census of Amherst from the census of 1911, see Jamieson, *Burma Gazetteer*, 17–18; Hallett, *A Thousand Miles on an Elephant*, 21.

59. Simmonds and Kingston, eds. *Colonial Magazine and East India Review* (1846): 195–97.

60. Suthep Soonthornpasuch, "Islamic Identity in Chiengmai City: A Historical and Structural Comparison of Two Communities" (PhD diss., University of California, 1977).

61. Soonthornpasuch, "Islamic Identity in Chiengmai City," 43–47; For another pioneering account of the Bengalis in Thailand, Cambodia, Annam, and Borneo in the pre–World War II period, see Gyanendramohon Das, *Banger Bahire Bangali*, part 3 (Indian Publishing House, 1931), 436–44.

62. Hallett, *A Thousand Miles on an Elephant*, 14, 36–37, 48, 46, 170.

63. "Pacification of the West Mang Lun," 7–11.

64. For example, the Pak Muang village on the Meh Kha. See Hallett, *A Thousand Miles on an Elephant*, 90.

65. "Pacification of the West Mang Lun," 7–11.

66. Hallett, *A Thousand Miles on an Elephant*, 174–75.

67. Prince Henri d'Orléans, *From Tonkin to India by the Sources of the Irawadi, January '95–January '96*, trans. Hamley Bent (Methuen & Co, 1898), 199.

68. C. H. Desgodins, *Mission du Thibet de 1855 à 1870: Relations et observations diverses sur les populations de l'Indo-Chine septentrionale, du Thibet, du pays des Lolos et des autres contrées de la Chine occidentale* (V. Palmén, 1872), 323; John Bray, "Trade, Territory, and Missionary Connections in the Sino-Tibetan Borderlands," in *Frontier Tibet: Patterns of Change in the Sino-Tibetan Borderlands*, ed. Stéphane Gros (Amsterdam University Press, 2019), 155.

69. Government of Burma, *Gazetteer of Upper Burma and the Shan States*, part 1, vol. 1 (Government of Burma, 1900), 12.

70. Ronald Kaulback, *Salween* (Hodder and Stoughton, 1938), 158; For a representative map of the fluvial system connected to Salween, see Coryton, "Trade Routes Between British Burmah and Western China," 229–30.

71. TBL, IOR/L/PS/533, collection 1: Journal of an expedition from Gyeen to Mandalay, via Karenne and the Shan States, kept by Captain C. E. Watson, Assistant commissioner, Martaban District, British Burma, April 20, 1864; See *Journal of the Salween Surveying Expedition, During the Season 1864–65* (Military Orphan Press, 1865), 15.

72. Kaulback, *Salween*, 159.

73. Ellen Thorp, *Quiet Skies on Salween* (Jonathan Cape, 1945), 70.

74. TBL, IOR/L/PS/533, collection 1: Journal of Lt. G. Colquhoun Sconce while employed on Salween Surveying expedition.

75. Thorp, *Quiet Skies on Salween*, 83

76. Thorp, *Quiet Skies on Salween*, 131.

77. Hallett, *A Thousand Miles on an Elephant*, 67.

78. Government of India, *Imperial Gazetteer of India*, 423.

Chapter 4

1. The Mekong is a short form of "Menam Kong," Menam being the generic name for "mother of waters" in Thai. The Mekong is known as Dzachu in Tibet, Lancang jiang in China, Lu Shan in Burma, Song Lon in Vietnam, and Tonle Thom in Cambodia, meaning "the great river." See Milton Osborne, *River Road to China: The Mekong River Expedition 1866–1873* (Readers Union, 1976), 39–44.

2. George Litton, *Report on a Journey in North and North-West Yunnan, Season 1902–1903* (Shanghai Mercury, 1903), 21.

3. Milton Osborne, *Mekong: Turbulent Past, Uncertain Future* (Allen & Unwin, 2006), 278.

4. TBL, IOR/L/PS/18: G.H. 19 Nov 1895: For a note on three distinct geomorphological segments of the Mekong across Laos and Cambodia, see Martin Stuart-Fox, *A History of Laos* (Cambridge University Press, 1997), 9.

5. Osborne, *Mekong*; Osborne, *River Road to China*.

6. Holt S. Hallett, "Western China," in *The Nineteenth Century: A Monthly Review* 38 (July–December 1895): 242; Fred W Carey, "A Trip to the Chinese Shan States," *The Geographical Journal* 14, no. 4 (October 1899): 395.

7. TBL, IOR/L/PS/18/B113: Trade of Szemao (Yunnan) with Burma. Its commercial importance. Mandalay-Kunlon Railway and its terminus on the Salween, 3.

8. Henry Rodolph Davies, *Yün-nan: The Link Between India and Yangtze* (Cambridge University Press, 1909), 96.

9. Prince Henri d'Orléans, *From Tonkin to India by the Sources of the Irawadi, January '95–January '96*, trans. Hamley Bent (Methuen & Co, 1898), 120.

10. R. Sprye "Communication with the South-West Provinces," 47; TBL, IOR/L/PS/18/B117: Memorandum on Indian Trade with Western China, and on the Resources, Trade, and Trade Route of Yunnan and Adjacent Provinces, 1899. 29.

11. Suthep Soonthornpasuch, "Islamic Identity in Chiengmai City: A Historical and Structural Comparison of Two Communities" (PhD diss., University of California, 1977), 10a.

12. Hallett, "Western China," 208, 210, 281; J. Coryton, "Trade Routes Between British Burmah and Western China," *Journal of the Royal Geographical Society of London* 45 (1875); see also Andrew Forbes and David Henley, *The Haw: Traders of the Golden Triangle* (Asia Film House, 1997).

13. Mekong River Commission, "Mekong Basin: Geography," https://www.mrcmekong.org/about/mekong-basin/geography/, accessed February 6, 2025.

14. George C. Hurlbut, "Geographical Notes," *Journal of the American Geographical Society of New York* 24 (1892): 118.

15. Lord Lamington, "Journey Through the Trans-Salwin Shan States to Tong-King," *Proceedings of the Royal Geographical Society and Monthly Record of Geography* 13, no. 12 (December 1891): 707.

16. Foreign Office, *Reports from Her Majesty's Consuls on the Manufactures, Commerce &tc.*, 13 (Harrison and Sons, 1873), 791–92.

17. Bennet Bronson, "Exchange at the Upstream and Downstream Ends: Notes Toward a Functional Model of the Coastal State in Southeast Asia," in *Economic Exchange and Social Interaction in Southeast Asia: Perspectives from Prehistory, History, and Ethnography*, ed. Karl Hutterer (University of Michigan Press, 1977), 39–52.

18. Oscar Salemink, "The Regional Centrality of Vietnam's Central Highlands," *Oxford Research Encyclopedia of Asian History* (Oxford University Press, 2018), https://doi.org/10.1093/acrefore/9780190277727.013.113.

19. Osborne, *River Road to China*, 220.

20. Stuart-Fox, *A History of Laos*, 16

21. For the complicated border relations between British Burma and Siam, see Thongchai Winichakul, *Siam Mapped: A History of the Geo-Body of a Nation* (University of Hawaii Press, 1997).

22. Peter Simms and Sanda Simms, *The Kingdoms of Laos: Six Hundred Years of History* (Curzon, 1999), 155.

23. TBL, IOR/L/PS/18: Memorandum of Communication with the French Government respecting Kyaing Cheng, 8–10.

24. Holly High, "Dreaming Beyond Borders: The Thai/Lao Borderlands and the Mobility of the Marginal," in *On the Borders of State Power: Frontiers in the Greater Mekong Sub-Region*, ed. Martin Gainsborough (Routledge, 2009), 84.

25. d'Orléans, *From Tonkin to India*, 121.

26. Osborne, *River Road to China*, 218–19.

27. Winichakul, *Siam Mapped*; Soo Mun Theresa Wong, "Making the Mekong: Nature, Region, Postcoloniality" (PhD Diss., The Ohio State University, 2010), 42.

28. TBL, IOR/L/PS/18: G. H. November 19, 1895.

29. Trade of Szemao (Yunnan) with Burma, 5.

30. Trade of Szemao (Yunnan) with Burma, 2.

31. High, "Dreaming Beyond Borders," 85.

32. Andrew Walker, "Conclusion: Are the Mekong Frontiers Sites of Exception?," in *On the Borders of State Power: Frontiers in the Greater Mekong Sub-Region*, ed. Martin Gainsborough (Routledge, 2009), 103.

33. Stuart-Fox, *A History of Laos*, 46.

34. OR/L/PS/18/B117: Memorandum on Indian Trade with Western China, and on the Resources, Trade, and Trade Route of Yunnan and Adjacent Provinces, (1899), 29.

35. Hallett, *A Thousand Miles*, 382.

36. Leo Alting Von Geusau, "Dialectics of Akhazang: The Interiorizations of a Perennial Minority Group," in *Highlanders of Thailand*, ed. John McKinnon and Wanat Bhruksasri (Oxford University Press, 1986), 253. This is similar to Tony Huber's notes on micromigration in the upper Brahmaputra network. See Huber, "Micro-Migrations of Hill Peoples in Northern Arunachal Pradesh: Rethinking Methodologies and Claims of Origins in Tibet," in *Origins and Migrations in the Extended Eastern Himalayas*, ed. Toni Huber and Stuart Blackburn (Brill, 2012).

37. For details on these messy political developments, see Simms and Simms, *The Kingdoms of Laos*, 181–83.

38. Litton, *Report on a Journey in North and North-West Yunnan*, 20–21.

39. David Biggs, *Quagmire: Nation-Building and Nature in the Mekong Delta* (University of Washington Press, 2020).

40. Philip Taylor, "Water in the Shaping and Unmaking of Khmer Identity on the Vietnam-Cambodia Frontier," *TRaNS: Trans-Regional and -National Studies of Southeast Asia* 2, no. 1 (2014): 103.

41. Mohamed Effendy Bin Abdul Hamid, "Understanding the Cham Identity in Mainland Southeast Asia: Contending Views," *SOJOURN: Journal of Social Issues in Southeast Asia* 21, no. 2 (2006): 232; See also Charles Wheeler, "One Region, Two Histories: Cham Precedents in the History of the Hôi An Region," in *Viet Nam: Borderless Histories*, ed. Nhung Tuyet Tran and Anthony Reid (University of Wisconsin Press, 2006), 71–75.

42. Hy Van Luong, "Mobile Trading Network from Central Coastal Vietnam," in *Traders in Motion: Identities and Contestations in the Vietnamese Marketplace*, ed. Kirsten W. Endres and Ann Marie Leshkowich (Cornell University Press, 2018), 95.

43. Christian Culas, "Migrants, Runaways and Opium Growers: Origins of the Hmong in Laos and Siam in the Nineteenth and Early Twentieth Centuries," in *Turbulent Times and Enduring Peoples: Mountain Minorities in the South-East Asian Massif*, ed. Jean Michaud and Jan Ovesen (Curzon, 2000), 37–39.

44. Chob Kacha-Ananda, "Migration, Settlements and Land," in *Highlanders of Thailand*, ed. John McKinnon and Wanat Bhruksasri (Oxford University Press, 1986), 212–13.

45. Robbins Burling, *Hill Farms and Padi Fields: Life in Mainland Southeast Asia* (Prentice-Hall, 1965), 57–58.

46. Karl Gustav Izikowitz, *Lamet: Hill Peasants in French Indochina* (AMS, 1979), 252.

47. Izikowitz, *Lamet*, 27.

48. Izikowitz, *Lamet*, 234, 274–75.

49. Andrew Walker, *The Legend of the Golden Boat: Regulation, Trade and Traders in the Borderlands of Laos, Thailand, China and Burma* (Curzon, 1999), 29.

50. Walker, *The Legend of the Golden Boat*, 195.

51. For fresh insights into the Lower Mekong trade network and transregional vibe, see Nola Cooke and Li Tana, eds., *Water Frontier: Commerce and the Chinese in the Lower Mekong Region, 1750–1880* (National University of Singapore Press, 2004).

52. Philip Taylor, "The Cosmopolitan Delta. Ethnic Pluralism at the Mouth of the Mekong," in *Routledge Handbook of Contemporary Vietnam*, ed. Jonathan D. London (Routledge, 2022).

53. Soonthornpasuch, "Islamic Identity in Chiengmai City," 49–54.

54. Soonthornpasuch, "Islamic Identity in Chiengmai City," 57.

55. Andrew D. W. Forbes, "The "ČĪN-HỌ̆" (Yunnanese Chinese) Caravan Trade with North Thailand During the Late Nineteenth and Early Twentieth Centuries", *Journal of Asian History* 21, no. 1 (1987): 13. Luang Prabang, at the great elbow of the Mekong, was one of those large *paak*, formed off the Mekong and Nam Khan Rivers.

56. Lord Lamington, "Journey Through the Trans-Salwin Shan States," 708.

57. Lamington, "Journey Through the Trans-Salwin Shan States," 14.

58. Litton, *Report on a Journey in North and North-West Yunnan*, 8.

59. R. S. Lyngdoh, *Ka Histori Ka Thoh ka Tar, bynta 1* (H. W. Sten, 1979), 9.

60. Walker, *The Legend of the Golden Boat*, 19.

61. Lamington, "Journey Through the Trans-Salwin Shan States," 707.

62. E. Paul Durrenberger, "Lisu: Political Form, Ideology and Economic Action," in *Highlanders of Thailand*, ed. John McKinnon and Wanat Bhruksasri (Oxford University Press, 1986), 215–26; For a note on the Lahu people, see Anthony R. Walker, "The Lahu People: An Introduction," in *Highlanders of Thailand*, 227–37.

63. Hallett, *A Thousand Miles*, 333.

64. Harold Jefferson Coolidge Jr. and Theodore Roosevelt, *Three Kingdoms of Indo-China* (Thomas Y. Crowell, 1933), 61, 149, 158.

65. Francis Kingdon Ward, *The Land of the Blue Poppy* (Cambridge University Press, 1913), 224.

66. Martin Heidegger, *Poetry, Language, Thought*, trans. Albert Hofstadter (Harper Colophon Books, 1971). For a recent Heideggerian perspective on dwelling on the Mekong as seen in a novel, see Chitra Sankaran, *Women, Subalterns and Ecologies in South and Southeast Asian Women's Fiction* (University of Georgia Press, 2021).

Chapter 5

1. In Vietnam the Red is known as Hồng Hà and Sông Cá (Fish River). In China, it is called Yüan Chiang (Original River). According to historian Li Tana (2016), the Red River is known as Nam Tao, meaning "Big River" or "Major River" among Tai-Kadai speakers. In Chinese historical records, the river is also referred to as Phu Luong and Song Lo, both terms being derived from the Austroasiatic language family and meaning "River." Its major tributaries, Black and Clear rivers, are called Song Da and Song Lo, respectively. The French called the river le Fleuve Rouge. The English name is used in this book, as it is most commonly found in literature covering this period.

2. Le Ba Thao, *Vietnam: The Country and its Geographical Regions* (Gioi, 2017), 301.

3. Nola Cooke, Li Tana, and James A. Anderson eds., *The Tongking Gulf Through History* (University of Pennsylvania Press, 2011).

4. Nanny Kim, *Mountain Rivers, Mountain Roads: Transport in Southwest China, 1700–1850* (Brill, 2020).

5. A. R. Agassiz, "From Hai-Phong in Tong-King to Canton, Overland," *Proceedings of the Royal Geographical Society and Monthly Record of Geography* 13, no. 5 (May 1891): 249–64.

6. Ulrich Theobald, "Southwest China: Local Conditions and Economic Trajectories," in *Southwest China in a Regional and Global Perspective (c.1600–1911): Money, Transport, Trade and Society*, ed. Ulrich Theobald and Cao Jin (Brill, 2018), 27–28.

7. Inspectorate General of Customs, *Decennial Reports on the Trade, Navigation, and Industries, etc., of the Ports Open to Foreign Commerce in China and Corea and on the Condition and Development of the Treaty Port Provinces 1882–91* (Statistical Department of the Inspectorate General of Customs, Shanghai, 1893), Appendix, x–xi, 667.

8. George Ernest Morrison, *An Australian in China* (Methuen & Co., 1895), 149.

9. Li Tana, "A Historical Sketch of the Landscape of the Red River Delta," *TRaNS: Trans-regional and -National Studies of Southeast Asia* 6 (2016): 359–60.

10. A. R. Colquhoun, "Tongquin," *Straits Times Weekly Issue* (October 20, 1883): 5.

11. Kuo Tsung-fei, "A Brief History of the Trade Routes Between Burma, Indochina and Yunnan," *T'ien Hsio Monthly* 12 (1941): 29.

12. TBL, IOR/L/PS/18/B117: Various Authorities, Memorandum on Indian Trade with Western China and On the Resources, Trade, and Trade Routes of Yunnan and Adjacent Provinces, 31.

13. F. v. Richthofen, "Recent Attempts to Find a Direct Trade-Road to South-Western China," *Oceanic Highways* (January 1874): 407–8. See also A. P. McMahon, "On Our Prospects of Opening a Route to South-Western China, and Explorations of the French in Tongquin and Cambodia," *Proceedings of the Royal Geographical Society of London* 18, no. 4 (1873–1874): 465.

14. Prince Henri d'Orléans, *From Tonkin to India by the Sources of the Irawadi, January '95–January '96*, trans. Hamley Bent (Methuen & Co, 1898), 13–14.

15. Alexander Hosie, *Three Years in Western China; A Narrative of Three Journeys in Ssu-ch'uan, Kuei-chow, and Yün-nan* (George Philip & Son, 1897), xvii.

16. Hosie, *Three Years in Western China*, xvi.

17. Other import items included clocks, watches, lamps, oil, glass crockery, trepan, aniline, dyes, ironmongery, magnets, musical boxes, wines and spirits, cigars, umbrellas, and fancy products, in addition to cotton.

18. Henry Rodolph Davies, *Yün-nan: The Link Between India and Yangtze* (Cambridge University Press, 1909), 178–79.

19. Colquhoun, "Tongquin," 7–8.

20. Various Authorities, Memorandum on Indian Trade with Western China, 31.

21. Hosie, *Three Years in Western China*, xvii.

22. Mss Eur F86/248: Letters from James Turner to John Halliday of the Anacan Company, London, October 20 and 30, 1897, with regard to the Yunnan country, 35.

23. J. Coryton and Mr. Margary, "Routes Between British Burmah and Western China; with Extracts of Letters from Mr. Margary," *Proceedings of the Royal Geographical Society of London* 19, no. 4 (1874–1875): 276.

24. Davies, *Yün-nan*, 178.

25. Richthofen, "Recent Attempts to Find a Direct Trade-Road," 410.

26. For a detailed exploration of this route, see d'Orléans, "A Journey from Tonkin by Tali-Fu to Assam," *The Geographical Journal* 8, no. 6 (December 1896): 566–82.

27. D'Orléans, *From Tonkin to India*, 86, 89.

28. Fred W. Carey, "A Trip to the Chinese Shan States," *The Geographical Journal* 14, no. 4 (October 1899): 388–90.

29. D'Orléans, "A Journey from Tonkin," 570.

30. Colquhoun, "Tongquin," 3.

31. Jean Michaud and Sarah Turner, "Tonkin's Uplands at the Turn of the 20th Century: Colonial Military Enclosure and Local Livelihood Effects," *Asia-Pacific View Point* 57, no. 2 (2016): 158–59.

32. Mss Eur C317/2: La Mission Doudart De Lagree – Francis Gariner (1866–1868) (Jean-Pierre GOMANE, 1976), 249; See also J. Dupuis, *A Journey to Yunnan and the Opening of the Red River to Trade*, trans. Walter E. J. Tips (originally published 1880; White Lotus, 1998), 87.

33. Kees van Dijk, *Pacific Strife* (University of Amsterdam, 2015), 277.

34. Mss Eur C317/2: La Mission Doudart De Lagree–Francis Garnier (1866–1868) (Jean-Pierre GOMANE, 1976), 249.

35. Morrison, *An Australian in China*, 43

36. Dupuis, *A Journey to Yunnan*, 4.

37. Richthofen, "Recent Attempts to Find a Direct Trade-Road," 406. See also McMahon, "On Our Prospects of Opening a Route to South-Western China," 464.

38. Jean Michaud, "French Missionary Expansion in Colonial Upper Tonkin," *Journal of Southeast Asian Studies* 35, no. 2 (June 2004): 301–3.

39. Michaud, "French Missionary Expansion," 305.

40. Jean Michaud, *"Incidental" Ethnographers: French Catholic Missions on the Tonkin-Yunnan Frontier, 1880s–1930* (Brill, 2007); See also Jean Michaud, "French Military Ethnography in Colonial Upper Tonkin (Northern Vietnam), 1897–1904," *Journal of Vietnamese Studies* 8, no. 4 (Fall 2013):1–46.

41. Letters from James Turner to John Halliday, 35.

42. Milton Osborne, *River Road to China: The Mekong River Expedition 1866–1873* (Readers Union, 1976), 39–44.

43. TBL, IOR/L/PS/20/D119: M. E. Willoughby, Captain of 2nd Bengal Lancers, Report on the Coast Defences of the Tonkin Delta, (Department of Government Printing, India, 1900), 4.

44. Colquhoun, "Tongquin," 3.

45. Archibald Little, *Across Yunnan and Tonking* (Chungking, 1904), 30; J. George Scott, "The Hill-Slopes of Tong-Kin," *Proceedings of the Royal Geographical Society and Monthly Record of Geography* 8, no. 4 (April 1886): 219.

46. Willoughby, Report on the Coast Defences of the Tonkin Delta, 3.

47. TBL, IOR/L/PS/18/B117: Various Authorities, Memorandum on Indian Trade with Western China, 32.

48. J. D. Mollon, "The Origins of the Concept of Interference," *Philosophical Transactions of the Royal Society London* 360, no. 1764 (2002).

49. Colquhoun, "Tongquin," 7–8.

50. Dupuis, *A Journey to Yunnan*, 41; Dupuis anchored at the mouth of the Black River (locally known as Tsong-po), from the name of a great village which was the endpoint of navigation for barges on the Red River (48).

51. TNA, WO 106/70: Report on the Coast Defences of the Tonkin Delta (1900), 1–2.

52. Dupuis, *A Journey to Yunnan*, 101.

53. Tsung-fei, "A Brief History of the Trade Routes," 29.

54. Howard Dick and Peter J. Rimmer, *Cities, Transport and Communications: The Integration of Southeast Asia since 1850* (Palgrave Macmillan, 2003), 62.

55. Tsung-fei, "A Brief History of the Trade Routes," 1, 31–32.

56. Little, *Across Yunnan and Tonking*, 12.

57. Little, *Across Yunnan and Tonking*, 26.

58. Carey, "A Trip to the Chinese Shan States," 380–85.

59. Pierre Brocheux and Daniel Hémery, *Indochina: An Ambiguous Colonization, 1858–1954* (University of California Press, 2011).

60. Ella S. Laffey, "French Adventures and Chinese Bandits in Tonkin: The Garnier Affair in Its Local Context," *Journal of Southeast Asian Studies* 6, no. 1 (March 1975): 39.

61. Laffey, "French Adventures and Chinese Bandits in Tonkin," 40, 46; for discussions on the Black Flags' predominance over the Yellow Flags on the Red River and the French lack of control on the strategic points of the Red River, see Osborne, *River Road to China*, 202, 205.

62. Peter Simms and Sanda Simms, *The Kingdoms of Laos: Six Hundred Years of History* (Curzon, 1999), 200; Colquhoun, "Tongquin," 8.

63. Colquhoun, "Tongquin," 2. Colquhoun calls these steps "the beginning of the end."

64. Philippe Le Failler, "The Dêo Family of Lai Châu: Traditional Power and Unconventional Practices," *Journal of Vietnamese Studies* 6, no. 2 (2011): 45–47.

65. Failler, "The Dêo Family of Lai Châu," 48.

66. d'Orléans, *From Tonkin to India*, 47, 59.

67. Philippe Le Failler, *La rivière Noire: L'intégration d'une marche frontière au Vietnam* (CNRS, 2019).

68. Inspectorate General of Customs. *Decennial Reports on the Trade, Navigation, and Industries*, 666.

69. Sarah Turner, "Borderlands and Border Narratives: A Longitudinal Study of Challenges and Opportunities for Local Traders Shaped by the Sino-Vietnamese

Border," *Journal of Global History* 5, no.2 (2010): 271; Sarah Turner and Jean Michaud, "'Weapons of the Weak': Selective Resistance and Agency among the Hmong in Northern Vietnam," in *Agrarian Angst and Rural Resistance in Contemporary Southeast Asia*, ed. Dominique Caouette and Sarah Turner (Routledge, 2009). For more on the Red River as a site of layered contestations, see Jean Michaud, ed., *Turbulent Times and Enduring Peoples: Mountain Minorities in the Southeast Asian Massif* (Curzon, 2000).

70. Bradley Camp Davis, *Imperial Bandits: Outlaws and Rebels in the China-Vietnam Borderlands* (University of Washington Press, 2017).

71. d'Orléans, *From Tonkin to India*, 10–11.

72. Veronica Strang, "Common Senses: Water, Sensory Experience and the Generation of Meaning," *Journal of Material Culture* 10, no. 1 (2005): 92–120.

73. d'Orléans, *From Tonkin to India*, 31.

74. Archibald John Little, *Across Yunnan: A Journey of Surprises* (Cambridge University Press, 1910), 109.

75. d'Orléans, *From Tonkin to India*, 34.

76. d'Orléans, "A Journey from Tonkin by Tali-Fu to Assam," *The Geographical Journal* 8, no. 6 (December 1896).

77. David Ludden, "Investing in Nature Around Sylhet," *Economic and Political Weekly* 38, no. 48 (2003): 5081.

78. Colquhoun, "Tongquin," 3.

79. Carey, *A Trip to the Chinese Shan States*, 384.

80. Christian C. Lentz, *Contested Territory: Ðien Biên Phu and the Making of Northwest Vietnam* (Yale University Press, 2019), 17, 29.

81. d'Orléans, 1898. *From Tonkin to India*, 73.

Chapter 6

1. The name Yangzi (Yangtze) is a Western misnomer derived from the river's lower stretch near Yangzhou. Historically, the river has had multiple regional names: Tang-Tze Kiang (Djre Kio) or Murui-ussu (Blue Water) in Tibet; Shingshakinag / Chinshakiang / Takiang (River of Golden Sand) in Western Sichuan. The Shingshakiang merges with the Min River at Sui-fu (modern Yibin) to form the Yangtzikiang, later known as Changjiang (Long River) or Da Jiang (Great River). See Keith Stevens, "A Tale of Sour Grapes: Messrs. Little and Mesny and the First Steamship Through the Yangzi Gorges," *Journal of the Hong Kong Branch of the Royal Asiatic Society* 41 (2001): 1.

2. Gretchen Mae Fitkin, *The Great River: The Story of a Voyage on the Yangtze Kiang* (North-China Daily News & Herald, 1922), iv. The figure of 770,000 cusecs at the mouth of the Yangzi is corroborated by the fact that about half a century earlier, the upper Yangzi, a few miles below Yichang, had a discharge of 500,000 cusecs. This "pretty sure" data was personally obtained through a sextant by Blakiston, a captain of the British Army. See Thomas W. Blakiston, *Five Months on the Yang-Tsze* (John Murray, 1862), 294–95.

3. For an authoritative note on the Yangzi's network of tributaries, see Mǎ Lìqín, *Chángjiāng huánghé* (National Publication Foundation, 2014), 2–26.

4. William Lockhart, "On the Importance of Opening the Navigation of the Yang-tse-Kiang, and the Changes That Have Lately Taken Place in the Bed of the Yellow River, &c.," *Proceedings of the Royal Geographical Society of London* 2, no. 4 (1857–1858): 202–3; For details of the premodern contexts of ecological changes in the Yellow River system, see Zhang, *The River, The Plain and the State*. On aspects of deterioration of the Yellow in the nineteenth century, see Charles K. Edmunds, "Some of China's Physical Problems," *The Journal of Race Development* 4, no. 2 (1913): 136–40; For recent notes on the advantages of the Yangzi over the Yellow in terms of fluvial ecology, agriculture, commerce, and shipping, as seen in Japanese and Chinese sources, see Fei Chen, "Rediscovering the Yellow River and the Yangtze River: The Circulation of Discourses on the North-South Dichotomy Between Late Qing China and Meiji Japan," *International Journal of Asian Studies* 16, no. 1 (2019): 45.

5. Lockhart, "On the Importance of Opening the Navigation of the Yang-tse-Kiang," 206.

6. William T. Rowe, "Economic Transition in the Nineteenth Century," in *The Cambridge Economic History of China*, ed. Debin Ma and Richard von Glahn (Cambridge University Press, 2022), 52.

7. TNA, 1941, PPMS2, Confidential Letters 1882–1941: Mr Maze's Speech on the Occasion of Laying the Foundation Stone of the Imposing new Hankow Customs House, November 4, 1922, 259.

8. TBL, IOR/L/PS/18/B117: Various Authorities, Memorandum on Indian Trade with Western China and On the Resources, Trade, and Trade Routes of Yunnan and Adjacent Provinces, 40.

9. Henry Rodolph Davies, *Yün-nan: The Link Between India and Yangtze* (Cambridge University Press, 1909), 8–9.

10. Archibald Duncan Blue, "The China Coast: A Study of British Shipping in Chinese Waters 1842–1914" (PhD diss., University of Strathclyde, 1982), 175; Anne Reinhardt, "Treaty Ports as Shipping Infrastructure," in *Treaty Ports in Modern China: Law, Land and Power*, ed. Robert Bickers and Isabella Jackson (Routledge, 2016), 107–8.

11. E. H. Wilson, *A Naturalist in Western China*, vol. 1 (Methuen & Company, 1913), 8–10.

12. William B. Dunlop, "The Key of Western China,", *Asiatic Quarterly Review* 7, (April 1889): 14.

13. Blue, "The China Coast," 180.

14. Alexander Hosie, *Three Years in Western China; A Narrative of Three Journeys in Ssu-ch'uan, Kuei-chow, and Yün-nan* (George Philip & Son, 1897), xii–xiii; Rumor had it among the Europeans that women in Sichuan even hatched the eggs of the silkworm in their breasts. See Anonymous, "Western China: Its Products and Trade," *Littell's Living Age* 186, no. 2409 (August 30, 1890).

15. Hosie, *Three Years in Western China*, 209; Blue, "The China Coast," 209.

16. Anonymous, "Western China: Its Products and Trade."

17. TBL, IOR/L/PS/18/B117: Various Authorities, Memorandum on Indian Trade with Western China, 39.

18. C. C. Manifold, "The Problem of the Upper Yang-Tze Provinces and Their Communications," *The Geographical Journal* 25, no. 6 (June 1905): 614.

19. Nanny Kim, *Mountain Rivers, Mountain Roads: Transport in Southwest China, 1700–1850* (Brill, 2020).

20. TBL, IOR/L/PS/6/564, part 152: Notes by T. T. Cooper on water communication between Hankow Chung-king and Chen-tu, the capital of Sz-chuen. Calcutta, February 20, 1869, 2–3.

21. TBL, IOR/L/PS/6/564 PARTS 152: Slip of Meeting of the Royal Geographical Society of 13th June, 1870, 5

22. "An Old Resident in Western China" noted to T. T. Cooper that located between the Mekong and Yangzi, and comprising nine great plains, Batang was in the middle of the "best water perhaps in the world, water rolling down from the top of an enormous mountain amidst some vast caverns of marble; good and numerous horses; strong little mules for weight to Bhamo; kind population." He advised that from Batang one could "easily communicate with Assam, Burmah, Dali, Kunming, Szechuan, Lassa, Pomi, and the gold mountains known as Kin-tchouan." See TBL, IOR/L/PS/6/564, part 152: Memorandum addressed to Mr. T. T. Cooper by an old resident in Western China. October 4, 1868.

23. T. T. Cooper, "Letter from Mr. T. T. Cooper, on the Course of the Tsan-Po and Irrawaddy and on Tibet," *Proceedings of the Royal Geographical Society of London* 13, no. 5 (1868–1869): 338.

24. Hosie, *Three Years in Western China*, 145; TBL, IOR/L/PS/18/B117: Various Authorities, Memorandum on Indian Trade with Western China, 40–41.

25. Thomas W. Blakiston, *Five Months on The Yang-Tsze* (John Murray, 1862), 129–30.

26. Avijeet Bhattacharya, *Journeys on the Silk Road Through Ages—Romance, Legend, Reality* (Zorba Books, 2017), 43.

27. George Litton, *Report on a Journey in North and North-West Yunnan, Season 1902–1903* (Shanghai Mercury, 1903), 15.

28. George Ernst Morrison, *An Australian in China: Being the Narrative of a Quiet Journey Across China to British Burma* (Horace Cox, 1895), 280.

29. TBL, Mss Eur F86/248: Letters from James Turner to John Halliday of the Aracan Company, London, October 20 and 30, 1897, with regard to the Yunnan Country, 4.

30. Inspectorate General of Customs, *Decennial Reports on the Trade, Navigation, and Industries, etc., of the Ports Open to Foreign Commerce in China and Corea and on the Condition and Development of the Treaty Port Provinces 1882–91* (Statistical Department of the Inspectorate General of Customs, Shanghai, 1893), vii appendix.

31. TBL, IOR/L/PS/18/B117: Various Authorities, Memorandum on Indian Trade with Western China, 16.

32. Iftekhar Iqbal, "From Zomia to Holon," *Suvannabhumi* 12, no. 2 (2020).

33. TBL, IOR/L/PS/18/B113: Trade of Szemao (Yunnan) with Burma. Its commercial importance. Mandalay-Kunlon Railway and its terminus on the Salween.

34. Anonymous, "Western China: Its Products and Trade," 525.

35. Anonymous, "Trade Between Burma and Western China," *Amrita Bazar Patrika*, August 28, 1912. For example, in 1921, an ounce of raw opium cost 25 cents at the site of production in Sichuan, 70 cents in Chongqing per tael weight, $1.20 in Ichang, $1.80 in Wuhan, and $3 in Shanghai. See Fitkin, *The Great River*, 141.

36. These developments are meticulously narrated in Kim, *Mountain Rivers*, 512.

37. Litton, *Report on a Journey in North and North-West Yunnan*, 1.

38. Morrison, *An Australian in China*, 149.

39. Wilson, *Naturalist in Western China*, 7; See also T. T. Cooper, "Expedition of Mr. T. T. Cooper from the Yang-tze-Kiang to Thibet and India," *Proceedings of the Royal Geographical Society of London* 12, no. 5 (1867–1868): 338.

40. G. William Skinner, "Regional Urbanization in Nineteenth-Century China," in *The City in Late Imperial China*, ed. G. William Skinner (Stanford University Press, 1977).

41. G William Skinner's "central-place" theory was presented in his article "Cities and the Hierarchy of Local Systems," in *The City in Late Imperial China*, ed. G. William Skinner (Stanford University Press, 1977), 282. For a recent critique of Skinner that shows how economic activities shattered the perceived boundary of macroregions, see Wang Di, *Kuachu fengbi de shijie: Changjiang shangyou quyu shehui yanjiu, 1644–1911* (Striking out of a closed world: A study on society in the *upper* Yangzi region, 1644–1911) (Zhonghua shuju, 1993).

42. E. B. Sladen, *Official Narrative of and Papers Connected with the Expedition to Explore the Trade Routes to China via Bhamo* (British Burma Press, 1879); Iftekhar Iqbal, "Reclaiming the Crossroads Between India and China: A View from the River," *Economic and Political Weekly* 49, no. 51 (2014).

43. Rutherford Alcock, *The Journey of Augustus Raymond Margary, From Shanghae to Bhamo, And Back to Manwyne* (Macmillan and Co., 1876), 98.

44. Archibald John Little, *Through the Yang-tse Gorges, Or, Trade and Travel in Western China* (1888; repr., Cambridge University Press, 2010), 5.

45. For a note on various English expeditions on the Yangzi valley, see A. D. Blue, "Land and River Routes to West China (With Especial Reference to the Upper Yangtze)," *Journal of the Hong Kong Branch of the Royal Asiatic Society* 16 (1976): 162–78.

46. TBL, IOR/L/PS/11/82: J. Jordan to Edward Gray, July 1914.

47. Blue, "The China Coast," 167.

48. TBL, Mss Eur F111/395: J. O. Miller, Private Secretary to the Viceroy, to L. W. Dane, Secretary to the Government of India in the Foreign Department, January 18, 1904.

49. Cord Eberspächer, *Die deutsche Yangtse-Patrouille. Deutsche Kanonenbootpolitik in China im Zeitalter des Imperialismus 1900–1914* (Verlag Dr. Dieter Winkler, 2004), 232. For German interest in the Yangzi, see also "Deutsche Interessen am Yangtze," *Ostasiatischer Lloyd* (August 17, 1900): 625–28.

50. Blue, "The China Coast," 195; IOR: L/PS/11/82: Manchester Trade of Commerce to Board of Trade, June 23, 1914.

51. TBL, IOR/L/PS/11/82: C Greene to Edward Gray, June 13, 1914.

52. TNA, FO 800-31: E Hoki to J. Jordan Peking, January 23, 1915, 217–18.

53. TBL, IOR/L/PS/11/82: Manchester Trade of Commerce to Board of Trade, June 23, 1914: Comparison of the British and Japanese Interests in the Yang-tsze Valley.

54. TBL, IOR/L/PS/11/81: Conyngham Greene to Edward Grey, June 12, 1914.

55. TNA, WO 106/5295: Hankow Under the Japanese. The Eclipse of British Interests in the Yangtsze Valley by a Hankow Resident, 1938.

56. TNA, FO-17-1174. Part 1 1890: K. B. Murray, Secretary, London Chambers of Commerce, to Marcuis of Salisbury, Secretary of State for Foreign Affairs, January 6, 1890, 4.

57. Blue, "The China Coast," 177.

58. Little, *Through the Yang-tse Gorges*, 17.

59. Scott Relyea, "Settling Authority: Sichuanese Farmers in Early Twentieth-Century Eastern Tibet," in *Frontier Tibet: Patterns of Change in the Sino-Tibetan Borderlands*, ed. Stéphane Gros (Amsterdam University Press, 2019), 3.

60. A British official elaborated: "They started with the plain words of a treaty, on their own admission, against them; the obnoxious clauses gone, and for it they have substituted one which places absolutely in their power the steam navigation of over 500 miles of one of the most valuable commercial waterways in the world. They may show blindness to their real interests; no doubt it does; but in the diplomatic game on this occasion there can be no question who is the victor." See TNA, FO-17-1174. Part 4, 1890: "The Chung King Agreement with China: Text of the Convention," *The Times*, September 16, 1890.

61. Anonymous, "Western China: Its Products and Trade."

62. Reinhardt, "Treaty Ports as Shipping Infrastructure," 107–8. For an excellent study of the limits of both China's sovereignty and the imperial powers on the Yangzi around steam navigation, see Anne Reinhardt, *Navigating Semi-Colonialism: Shipping, Sovereignty, and Nation-Building in China, 1860–1937* (Harvard University Press, 2018).

63. For a note on the Faraizi activities in the lower Ganga-Brahmaputra, see Iftekhar Iqbal, *The Bengal Delta: Ecology, State and Social Change 1840–1943* (Palgrave, 2010), 67–93.

64. For notes on the Yangzi as the main site of Taiping rebellion, see Tobie Meyer-Fong, *What Remains: Coming to Terms with Civil War in 19th Century China* (Stanford University Press, 2013), 1, 5. See also Stephen Platt, *Autumn in the Heavenly Kingdom: China, the West, and the Epic Story of the Taiping Civil War* (Alfred A. Knopf, 2012), 39, 51.

65. Platt, *Autumn in the Heavenly Kingdom*, 62–63.

66. Platt, *Autumn in the Heavenly Kingdom*, 219, 27, 153–54.

67. David G. Atwill, *Chinese Sultanate: Islam, Ethnicity and the Panthay Rebellion in Southwestern China* (Stanford University Press, 2006).

68. Blakiston, *Five Months on the Yang-Tsze*, 121.

69. William Percival, *The Land of the Dragon: My Boating and Shooting Excursions to the Gorges of Upper Yangtze* (Hurst and Blackett, 1889), 96. The map accompanying Sarel's book also includes a place named "Mahomat" further upstream beyond Mussulman Point located by the Yichang Gorge. See Henry Andrew Sarel, "Notes on the Yang-tsze-Kiang, from Han-kow to Ping-shan," *The Journal of the Royal Geographical Society of London* 32 (1862): 2.

70. Blakiston, *Five Months on the Yang-Tsze*, 251. In response to my query, Jianxiong Ma, a colleague from Yunnan with expertise in the region, noted that he was unsure of the exact location of "Mussulman Point" in Yichang. However, he suggested that it could refer to certain streets within the city or nearby villages, as Hui Muslim settlements were common along the Yangzi River. He further noted the presence of Hui residential districts along the tributaries of the Yangzi from Yunnan to Sichuan, as well as a small Hui community in Yibin. Yibin, he added, serves as a port city near Zhaotong, a location he had previously visited. Jianxiong Ma, email message to author, February 14, 2022.

71. TBL, IOR/L/PS/6/564, part 152: Notes by T. T. Cooper on water communication between Hankow Chung-king and Chen-tu.

72. TNA, CIM/PP 482, part 5, Reel 102, 1919–1923: Additional Personal Papers of Geraldine Howard Taylor: China Inland Mission, 1865–1951, 40–41.

73. Noted in a review, by an unknown author, of Prince Henri d'Orléans's *Atour du Tonkin*, *The Edinburgh Review* (January 1896): 241–42.

74. Letters from James Turner to John Halliday, 42.

75. Arthur Davenport, *Report by Mr Davenport upon the Trading Capabilities of the Country Traversed by the Yunnan Mission* (Harrison and Sons, 1877), 11.

76. Litton, *Report on a Journey in North and North-West Yunnan*, 13.

77. Thomas M. Ainscough, *Notes From a Frontier* (Kelly and Walsh, 1915), 9.

78. TBL, IOR/L/PS/11/15: From a correspondent at Chengtu, Western Szechuan and the Republic. The Chia Kung States, *The Times*, July 21, 1913.

79. Reinhardt, *Navigating Semi-Colonialism*, 126

80. TNA, PREM11/2300 160 and 161: Washington Telegram no. 2462 to Foreign Office.

81. Blakiston, *Five Months on the Yang-Tsze*, 238.

82. TNA, FO-17-1174. Part 2, 1890: quoted in William B. Dunlop, "The Key of Western China," 7.

83. Blue, "The China Coast," 173, 177, 209.

84. Blakiston, *Five Months on the Yang-Tsze*, 238.

85. Inspectorate General of Customs, *Decennial Reports on the Trade, Navigation, and Industries*, 113–15.

86. Various Authorities, Memorandum on Indian Trade with Western China, 38.

87. Sarel, "Notes on the Yang-tsze-Kiang," 7; Thomas W. Blakiston, *The Yang-Tsze; with a Narrative of the Exploration of Its Upper Waters, and Notices of the Present Rebellions in China* (John Murray, 1862), 107.

88. Kim, *Mountain Rivers*, 489.

89. It is not entirely out of place to recall the Southern Tibetan perception of the boatman as "more beautiful than a god." See Diana Lange, "'The Boat Man Is More Beautiful than a God': Poetizing and Singing on the Rivers in Central and Southern Tibet," in *The Illuminating Mirror: Tibetan Studies in Honour of Per K. Soerensen on the Occasion of His 65th Birthday*, ed. Guntram Hazod and Olaf Czaja (Ludwig Reichert Verlag, 2016), 269–82.

90. Igor Iwo Chabrowski, "'Tied to a Boat by the Sound of a Gong': World, Work and Society Seen Through the Work Songs of Sichuan Boatmen (1880–1930s)" (PhD diss., European University Institute, 2013), 153; See also Yiying Pan, "Nexus of Self-Organization: The Expansion of Collective Responsibility Networks Among Boatmen in Nineteenth-Century Chongqing," *International Journal of Asian Studies* 20, no. 1 (2023).

91. Blakiston, *Five Months on The Yang-Tsze*, 120; For instances of invocation of Buddha while passing the rapids and gorges, see Davenport, *Report by Mr Davenport*, 9.

92. Little, *Through the Yang-tse Gorges*, 106–7, 132.

93. Blakiston, *Five Months on the Yang-Tsze*, 130.

94. TNA, FO-17-1174. Part 2, 1890: William B. Dunlop, "The Key of Western China," *Asiatic Quarterly Review* (April 1889): 5.

95. Alcock, *The Journey of Augustus Raymond Margary*, 100–101.

96. Morrison, *An Australian in China*, 73.

97. Blakiston, *Five Months on the Yang-Tsze*, 116, 231, 315.

98. Tansen Sen, *India, China and the World: A Connected History* (Rowman and Littlefield, 2017).

Chapter 7

1. TBL, IOR/L/PS/18/B117: Memorandum on Indian Trade with Western China and on the Resources, Trade, and Trade Routes of Yunnan and Adjacent Provinces, 1899, 44; The similitude of "sea of rugged mountains" comes from Archibald Little, *Across Yunnan: A Journey of Surprises* (Sampson Low, 1910), 87.

2. Emma Reisz, "Projecting the Road: Topological Photography on the Yunnan-Burma Frontier," *The Chinese Historical Review* 25, no. 2 (2018): 158–59.

3. Alexander Hosie, *Three Years in Western China; A Narrative of Three Journeys in Ssu-ch'uan, Kuei-chow, and Yün-nan* (George Philip & Son, 1897), 221.

4. William G. Clarence-Smith, "Breeding and Power in Southeast Asia: Horses, Mules and Donkeys in the Longue Durée," in *Environment, Trade and Society in Southeast Asia Book*, ed. David Henley and Henk Schulte Nordholt (Brill, 2015), 35.

5. TBL, IOR/L/MIL/7/1061 Collection 14/59: M. E. Willoughby, *Report on the Mules and Ponies*, 11–13.

6. Christian Daniels and Jianxiong Ma, "Introduction: The Agency of Local Elites in the Transformation of Western Yunnan During the Ming Dynasty," in *The

Transformation of Yunnan in Ming China: From the Dali Kingdom to Imperial Province, ed. Christian Daniels and Jianxiong Ma (Routledge, 2020), 5.

7. TNA, Annual Trade Reports 1900–1928: Chinese Maritime Customs Service, vol. I, 36.

8. The wartime needs of the mules as transporter were extensively discussed in TBL, IOR/L/PS/12/46/22: Notes on possibility of bringing supplies to China by pack traffic, 1942.

9. TBL, Mss Eur F86/248: Letters from James Turner to John Halliday of the Aracan Company, London, October 20 and 30, 1897, about the Yunnan country, 3, 28. For detailed notes on multiple commercial routes, see Nanny Kim, *Mountain Rivers, Mountain Roads: Transport in Southwest China, 1700–1850* (Brill, 2020).

10. Carol Ann Benedict, *Bubonic Plague in Nineteenth-Century China* (Stanford University Press, 1996), 28; C. Patterson Giersch, *Asian Borderlands: The Transformation of Qing China's Yunnan Frontier* (Harvard University Press, 2006), 170–71.

11. Andrew D. W. Forbes, "The "ČİN-HǪ" (Yunnanese Chinese) Caravan Trade with North Thailand During the Late Nineteenth and Early Twentieth Centuries," *Journal of Asian History* 21, no. 1 (1987): 10, 14.

12. Ch'ao-ting Chi, *Wartime Economic Development of China* (Garland, 1980)

13. Forbes, "The "ČİN-HǪ," 7, 10fn.

14. Memorandum on Indian Trade with Western China, 23.

15. TBL, Mss Eur F278/71: Report on the Shan States, 1890, xii.

16. Benedict, *Bubonic Plague*, 28; George E. Morrison, *An Australian in China, Being the Narrative of a Quiet Journey Across China to British Burma* (Horace Cox, 1895), 149.

17. Memorandum on Indian Trade with Western China, 28.

18. Forbes, "The "ČİN-HǪ," 17.

19. Holt S. Hallett, *A Thousand Miles on an Elephant in the Shan States* (William Blackwood and Sons, 1890), 172–73.

20. Generally, mules numbering more than five hundred were dealt with by the Hui or the Yunnan Muslims. For example, the firm (Hsing Shun Ho in Yunnan-fu) of Ma Taotai, a major Muslim businessman from Yunnan, traded annually with Burma and Siam with a caravan of five hundred mules and ponies, one half going down from Pu'er and Simao to Chiang Mai (Zimme) while the other half went westward through the southern Shan States in Mandalay. See Willoughby, *Report on the Mules and Ponies*, 20.

21. H. R. Davies, *Yün-nan: The Link Between India and Yangtze* (Cambridge University Press, 1909), 2.

22. TBL, IOR/L/PS/12/4622: A. G. N. Ogdon, Consul General, British Consulate, Kunming to Ambassador to China, British Embassy Chungqing, April 25, 1945.

23. F. KingdonvWard, *In Farthest Burma* (Seeley, Service & Co., 1921), 253.

24. John L. Christian, "Trans-Burma Trade Routes to China," *Pacific Affairs* 13, no. 2 (June 1940): 173–74.

25. Kuo Tsung-fei, "A Brief History of the Trade Routes Between Burma, Indochina and Yunnan." *T'ien Hsio Monthly* 12, no. 1 (1941): 30.

26. Fernand Braudel, *The Mediterranean and the Mediterranean World in the Age of Philip II*, vol. 1 (University of California Press, 1995), 317, 220.

27. Wen-Chin Chang, *Beyond Borders: Stories of Yunannese Chinese Migrants of Burma* (Cornell University Press, 2014), 152.

28. Letters from James Turner to John Halliday, 3.

29. George Litton, *Report on A Journey in North and North-West Yunnan, Season 1902–1903* (Shanghai Mercury, 1903), 21.

30. Prince Henri d'Orléans, "A Journey from Tonkin by Tali-Fu to Assam," *The Geographical Journal* 8, no. 6 (December 1896): 569.

31. G. C. Hurlbut, "Geographical Notes," *Journal of the American Geographical Society of New York* 24 (1892): 94–141.

32. Mss Eur F86/248: Letters from James Turner to John Halliday.

33. Letters from James Turner to John Halliday.

34. Benedict, *Bubonic Plague in Nineteenth-Century China*.

35. Quoted in Forbes "The "ČĪN-HỌ̆," 29; Willoughby noted the mule's comfortable encounter with "mere rock stairways and in other knee-deep quagmires of holding clay." Willoughby, *Report on the Mules and Ponies*, 6.

36. Ronald Kaulback, *Salween* (Hodder and Stoughton, 1938), 161.

37. Forbes, "The "ČĪN-HỌ̆," 4.

38. Willoughby, *Report on the Mules and Ponies*, 6.

39. Willoughby, *Report on the Mules and Ponies*, 4.

40. Prince Henri d'Orléans, *From Tonkin to India by the Sources of the Irawadi, January '95–January '96*, trans. Hamley Bent (Methuen & Co., 1898), 18.

41. Prince Henri d'Orléans, "From Yun-nan to British India," *The Geographical Journal* 7, no. 3 (March 1896): 301.

42. Willoughby, *Report on the Mules and Ponies*, 2.

43. Report on the Shan States, 1890, 9–10.

44. Nanny Kim, "Fuel for the Smelters: Copper Mining and Deforestation in Northeastern Yunnan During the High Qing, 1700 to 1850," in *Southwest China in a Regional and Global Perspective (c.1600–1911)*, ed. Ulrich Theobold and Cao Jin (Brill, 2018), 121.

45. Willoughby, *Report on the Mules and Ponies*, 6.

46. TNA, Pack animals. PPMS2, confidential letters, 1882–1941: Commissioner, Custom House, Tengyueh to Inspector General of Customs, Peking, July 29, 1907, 76.

47. TBL, IOR/L/PS/12/4622: H. J. Seymour, British Embassy Chungking, to Secretary to the Government of India in the External Affairs Department, New Delhi, March 18, 1942.

48. Willoughby, *Report on the Mules and Ponies*, 22, 29.

49. Willoughby, *Report on the Mules and Ponies*, 5–6. Hallett, *A Thousand Miles on an Elephant*, 212.

50. d'Orléans "From Yun-nan to British India," 301. See also Report on the Shan States, 1890.

51. Willoughby, *Report on the Mules and Ponies*, 8.

52. d'Orléans, *From Tonkin to India*, 52.

53. TBL, IOR/L/PS/12/4622: M. C. Gillet, Consul, H.B.M. Consulate, Tengyueh, camp Yunnafu to H.M. Embassy, Chungking, April 3, 1942.

54. James L. Hevia, *Animal Labour and Colonial Warfare* (University of Chicago Press, 2018).

55. Recent researches have highlighted the role of muleteers and their networks as key catalysts in the process of translocal commercial mobility. See C. Patterson Giersch, "Across 'Zomia' with Merchants, Monks, and Musk: Process Geographies, Trade Networks, and the Inner-East-Southeast Asian Borderlands," *Journal of Global History* 5, no. 2 (2010): 215–39; Peter C. Perdue, "Crossing Borders in Imperial China," in *Asia Inside Out: Connected Places*, ed. Eric Tagliacozzo, Helen F. Siu, and Peter C. Perdue (Harvard University Press, 2015).

56. Arthur Davenport, *Report by Mr Davenport upon the Trading Capabilities of the Country Traversed by the Yunnan Mission* (Foreign Office, 1877), 23.

57. Hosie, *Three Years in Western China*, 55.

58. Ann Maxwell Hill, *Merchants and Migrants: Ethnicity and Trade Among Yunnanese Chinese in Southeast Asia* (Yale University Press, 1998).

59. Forbes, "The "ČĪN-HỌ̈," 2, 4, 7.

60. Jean Berlie, "Links Between Muslims in Yunnan and Northern Thailand," in *Where China Meets Southeast Asia: Social and Cultural Change in the Border Regions*, ed. Grant Evans, Christopher Hutton, and Kuah Khun Eng (Institute of Southeast Asian Studies, 2000), 226, 234. David Atwill offers a detailed note on Yunnan's Muslim network in *Chinese Sultanate: Islam, Ethnicity, and the Panthay Rebellion in Southwest China, 1856–1873* (Stanford University Press, 2006), 43–48.

61. Ma Jianxiong and Ma Cunzhao, "The Mule Caravans of Western Yunnan: An Oral History of the Muleteers of Zhaozhou," *Transfers* 4, no. 3 (Winter 2014): 24–42; Jianxiong Ma and Cunzhao Ma, "The Mule Caravans as Cross-Border Networks: Local Bands and their Stretch on the Frontier Between Yunnan and Burma," in *Myanmar's Mountain and Maritime Borderscapes: Local Practices, Boundary-Making and Figured Worlds*, ed. Su-Ann Ho (ISEAS-Yusof Ishak Institute, 2016).

62. Jean Michaud and Christian Culas, "The Hmong of the Southeast Asia Massif: Their Recent History of Migration," in *Where China Meets Southeast Asia: Social and Cultural Change in the Border Regions*, ed. Grant Evans, Christopher Hutton, and Kuah Khun Eng (Institute of Southeast Asian Studies, 2000), 103–4.

63. Mika Toyota, "Cross-Border Mobility and Social Networks: Akha Caravan Traders," in *Where China Meets Southeast Asia: Social and Cultural Change in the Border Regions*, ed. Grant Evans, Christopher Hutton, and Kuah Khun Eng (Institute of Southeast Asian Studies, 2000), 207.

64. Betti Rosita Sari, "Trade Activities in Thailand Border Area Case Study: Mae Sai and Chiang Khong," in *Borders and Beyond: Transnational Migration and*

Diaspora in Northern Thailand Border Areas with Myanmar and Laos, ed. Betti Rosita Sari (Yayasan Pustaka Obor Indonesia, 2018), 170.

65. Andrew Forbes and David Henley, *Traders of Golden Triangle* (Cognoscenti Books, 2013).

66. Quoted in Wen-Chin Chang, "Circulations via Tangyan, a Town in the Northern Shan State of Burma," in *Asia Inside Out: Connected Places*, ed. Eric Tagliacozzo, Helen F. Sue, and Peter Perdue (Harvard University Press, 2015), 246–47.

67. Karl Gustav Izikowitz, *Lamet: Hill Peasants in French IndoChina* (AMS Press, 1979), 32, 65. This resembles the Zayat along the Salween basin.

68. Izikowitz, *Lamet*. 314.

69. Ann Maxwell Hill, "The Yunnanese: Overland Chinese in Northern Thailand," in *Chinese Society in Thailand: From Rural Markets to Global Markets*, ed. John McKinnon and Wanat Bhruksasri (Oxford University Press, 1986), 126.

70. Hill, "The Yunnanese," 123–34.

71. Moshe Yegar, *The Muslims of Burma: A Study of a Minority Group* (Verlag Otto Harrassowitz, 1972), 46.

72. Françoise Pommaret, "Ancient Trade Partners: Bhutan, Cooch Bihar, and Assam (17th–19th centuries)," *Journal of Bhutan Studies* 2, no. 1 (Autumn 2000): 9. https://lib.icimod.org/record/10468. Accessed February 5, 2025. See also Joy Pachuau, *Entangled Lives: Human-Animal-Plant Histories of the Eastern Himalayan Triangle* (Cambridge University Press, 2022), 57.

73. Hallett, *A Thousand Miles on an Elephant*, 196.

74. Suthep Soonthornpasuch, "Islamic Identity in Chiengmai City: A Historical and Structural Comparison of Two Communities" (PhD diss., University of California, 1977), 53–54.

75. Chang-Kuan Lin, "Chinese Muslims of Yunnan, Southwest China, with Special Reference to Their Revolt, 1855–1873" (PhD diss., University of Aberdeen, 1991), 85–86, 276.

76. Jianxiong Ma and Cunzhao Ma, "The Mule Caravans of Western Yunnan," 215–38.

77. John Leonard, "Books of the Times," *The New York Times*, May 20, 1982.

78. For a recent study of the location of the animal vis-à-vis the empire and the subalterns in Myanmar, see Jonathan Saha, *Colonizing Animals: Interspecies Empire in Myanmar* (Cambridge University Press, 2022).

79. Donna J. Haraway, *Staying with the Trouble: Making Kin in the Chthulucene* (Duke University Press, 2016); Anna Tsing et al., *Feral Atlas* (Stanford University Press, 2020), https://feralatlas.org. For an exploration of yak personhood in the context of animal-human relationship, see Jelle Wouters, "Relatedness, Trans-species Knots and Yak Personhood in Bhutan Highlands," in *Environmental Humanities in the New Himalayas: Symbiotic Indigeneity, Commoning and Sustainability*, ed. Dan Smyer Yü and Erik de Maaker (Routledge, 2021).

80. Jocelyne Porcher, "Animal Work," in *A Cultural History of Animals*, vol. 5, *In the Age of Empire*, edited by Kathleen Kete (Bloomsbury, 2011). For a brilliant study of the

historical implications of the place and displacement of animals in Ottoman Egypt, see Alan Mikhail, *The Animal in Ottoman Egypt* (Oxford University Press, 2014).

81. Little, *Across Yunnan*, 87.

Reflections

1. Richard White, *Organic Machine: The Making of the Columbian River* (Hill and Wang, 1996).

2. For an important collection of essays in this area, see E. Summerson Carr and Michael Lempert, eds., *Scale: Discourse and Dimensions of Social Life* (University of Chicago Press, 2016).

3. Quoted in J. Chandran, "The Burma-Yunnan Railway: Anglo-French Rivalry in Mainland Southeast Asia and South China, 1895–1902," (Ohio University Center for International Studies, 1971), 96. For Curzon's similar attitude against the railway in this region, including the one between Lashio and the Kunlon on the Salween, see Mss Eur. F111/536: The Foreign Department, vol. 5, 8–9; Mss Eur F86/248: Letters from James Turner to John Halliday of the Aracan Company, London, October 20 and 30, 1897.

4. John Percy Hardiman, *Gazetteer of Upper Burma and the Shan States*, vol. 1 (Government of Burma, 1900), 4.

5. Nurit Bird-David, "'Animism' Revisited: Personhood, Environment, and Relational Epistemology." *Current Anthropology* 40 (Supplement, February 1999); Veronica Strang, "Fluid Consistencies: Material Relationality in Human Engagement with Water," *Anthropological Dialogues* 21, no. 2; 133–50.

6. For example, in Yunnan hydroelectric project undertaken by the Bergstrom and Young company (On behalf of Metro-Vickers and other British electrical concerns) in 1943 at Hsiakwan couldn't be implemented due to lack of railway communication. See TBL, IOR/L/PS/12/4622: A. G. N. Ogdon, Consul General, British Consulate, Kunming to Ambassador to China, British Embassy Chungqing, April 25, 1945.

7. TNA, PPMS2, confidential letters,1882–1941: Memo on the Question of a "Yangtze Conservancy Board." March 20, 1920, 239.

8. Iftekhar Iqbal, "Locating the Riparian Commons in Eastern South Asia: A Translocal Perspective," in *Urban Development and Environmental History in Modern South Asia*, ed. Ian Talbot and Amit Ranjan (Routledge, 2022).

9. For a perceptive discussion of the significance of the Tibetan-Himalaya uplands and their anthropocenic implications, see Sunil Amrith and Dan Smyer Yü, "The Himalaya and Monsoon Asia: Anthropocenic Climes Since the 1800s," in *Storying Multipolar Climes of the Himalaya, Andes and Arctic: Anthropocenic Climate and Shapeshifting Watery Lifeworlds*, ed. Dan Smyer Yü and Jelle P. Wouters (Routledge, 2023), 29–51.

10. Vandana Shiva, *Water Wars* (Pluto, 2002); Iqbal, "Locating the Riparian Commons in Eastern South Asia," 106.

11. U.S. Geological Survey, "EarthView—Three Gorges Dam Brings Power, Concerns to Central China," November 17, 2016, https://www.usgs.gov/news/science-snippet/earthview-three-gorges-dam-brings-power-concerns-central-china; Yaxiang Wang et al., "Influence of Three Gorges Dam on Earthquakes Based on GRACE Gravity Field," *Open Geosciences* 14, no. 1 (2022): 453–61, https://doi.org/10.1515/geo-2022-0350, accessed February 6, 2025.

12. Harish Gupta, Shu-Ji Kao, and Minhan Dai, "The Role of Mega Dams in Reducing Sediment Fluxes: A Case Study of Large Asian Rivers," *Journal of Hydrology* 464–65 (September 25, 2012). https://doi.org/10.1016/j.jhydrol.2012.07.038, accessed February 5, 2025.

13. NJU Xinjinzhe, "Two Years After Yangtze Ban, Its Fishers Are Still Reeling," *Sixth Tone*, November 27, 2021, https://www.sixthtone.com/news/1009065#:~:text=According%20to%20the%20Yangtze%20River,roughly%20halving%20in%20the%201990s, accessed February 5, 2025.

14. Andrew Alan Johnson, *Mekong Dreaming: Life and Death Along a Changing River* (Duke University Press, 2020), 1.

15. Michael E. Webber, *The Thirst for Power. Energy, Water, and Human Survival* (Yale University Press, 2016), 37–38; See also Kenneth Pomeranz, "The Great Himalayan Watershed: Water Shortages, Mega-Projects and Environmental Politics in China, India, and Southeast Asia," *The Asia-Pacific Journal* 7, no. 30 (July 2009).

16. See Selina Ho, David M. Lampton, and Cheng-Chwee Kuik, *Rivers of Iron: Railroads and Chinese Power in Southeast Asia* (University of California Press 2020); Elisabeth Kaske, "Sichuan as a Pivot: Provincial Politics and Gentry Power in Late Qing Railway Projects in Southwestern China", in *Southwest China in a Regional and Global Perspective (c.1600–1911): Money, Transport, Trade and Society*, ed. Ulrich Theobald and Cao Jin (Brill, 2018), 416.

17. Chris Alden and Elizabeth Sidiropolos, "Silk, Cinnamon and Cotton: Emerging Power Strategies for the Indian Ocean and the Implications," South African Institute of International Affairs, *Policy Insights* 18 (June 2015); Joydeep Thakur, 'India Aims to Revive "Cotton Route" to Counter Chinese Silk Route', *Hindustan Times*, March 26, 2015, 4. https://iora.ris.org.in/sites/default/files/Hindustan_Times_Kolkata2015-03-26_page4.pdf

18. "The Tibetan Plateau Water Stores Under Threat," https://www.straitstimes.com/asia/east-asia/tibetan-plateau-water-stores-under-threat-study, accessed January 19, 2025. See also Ryan Woo, "It's So Hot in China, Melting Glaciers Risk Collapsing Dams,", *Sunday Morning Herald*, July 23, 2022, https://www.smh.com.au/world/asia/it-s-so-hot-in-china-melting-glaciers-risk-collapsing-dams-20220723-p5b3y2.html, accessed February 6, 2025; Jorge Daniel Taillant, *Glaciers: The Politics of Ice* (Oxford University Press, 2015), 153, 159–60. For the implication of global warming on the more than 2 billion people dependent on the BISMRY rivers, see Michael Buckley, *Meltdown in Tibet* (Palgrave Macmillan, 2014).

19. Qiang Zhang et al., "Oceanic Climate Changes Threaten the Sustainability of Asia's Water Tower," *Nature* 615 (March 2023): 87–93, https://doi.org/10.1038/s41586-022-05643-8, accessed January 19, 2025.

20. For an outline of the "corruption of memory" of intimate human-river relations under the condition of the adaptation regime, see Naveeda Khan, "The River and the Corruption of Memory," *Contributions to Indian Sociology* 49, no. 3 (2015): 389–409.

21. Iqbal, "Locating the Riparian Commons in Eastern South Asia."

22. Altogether close to four hundred dams have been built on the Brahmaputra up to 2020. See Paul F. Hudson, *Flooding and Management of Large Fluvial Lowlands: A Global Environmental Perspective* (Cambridge University Press, 2021), 88.

23. Sadiqur Rahman, "Is It Possible to Artificially Narrow the Jamuna River?" *The Business Standard*, March 22, 2023, https://www.tbsnews.net/features/panorama/it-possible-artificially-narrow-jamuna-river-603494, accessed February 5, 2025

24. David Blackbourn, "Time Is a Violent Torrent: Constructing and Reconstructing Rivers in Modern German History," in *Rivers in History: Perspectives in Europe and North America*, ed. Christof Mauch and Thomas Zeller (University of Pittsburgh Press, 2008).

25. S. G. Burrard and H. H. Hayden, *A Sketch of the Geography and Geology of the Himalaya Mountains and Tibet, part III, The Rivers of the Himalay and Tibet* (Superintendent Government Press, 1907), 126. Such a massive reduction in water discharge is likely true for the rest of the BISMRY rivers as well.

26. For historically informed explorations of contemporary predicaments of the Kachins, see Mandy Sadan, *Being and Becoming Kachin: Histories Beyond the State in the Borderworlds of Burma* (Oxford University Press, 2013); Karin Dean, *The Kachin Tackling the Territorial Trap: A Nation Divided by the Sino-Myanmar Boundary* (VDM Publishing, 2010).

27. Carl Middleton and Vanessa Lamb, eds., *Knowing the Salween River: Resource Politics of a Contested Transboundary River* (Springer Open, 2019), 34; see also Jerome Whitington, *Anthropogenic Rivers: The Production of Uncertainty in Lao Hydropower* (Cornell University Press, 2019).

28. Brian Eyler and Regan Kwan, "All Dams Map of the Mekong Basin: Mainstream and Tributaries," The Stimson Center, May 7, 2024, https://www.stimson.org/2024/all-dams-map-of-the-mekong-basin/, accessed February 5, 2025.

29. Brian Eyler, *Last Days of the Mighty Mekong* (Zed Books, 2019); For more recent developments on the Mekong, see also Nga Dao and Anh Tuan Le, eds., *Vietnam Hydropower and Its Challenges to Sustainability* (Science and Technology Publisher, 2016); Pamela McElwee, *Forests Are Gold: Trees, People, and Environmental Rule in Vietnam* (University of Washington Press, 2016).

30. Soo Mun Theresa Wong, "Making the Mekong: Nature, Region, Postcoloniality" (PhD diss., Ohio State University), 4.

31. Tana Li, "A Historical Sketch of the Landscape of the Red River Delta," *TRaNS: Trans-Regional and -National Studies of Southeast Asia* 4, no. 2 (July 2016): 352;

Tana Li, "Towards an Environmental History of the Eastern Red River Delta, Vietnam, c. 900–1400," *Journal of Southeast Asian Studies* 45, no. 3 (October 2014).

32. Sarah Turner, "Slow Forms of Infrastructural Violence: The Case of Vietnam's Mountainous Northern Borderlands," *Geoforum* 133 (July 2022): 189–97.

33. Philippe Le Failler, *La rivière Noire: L'intégration d'une marche frontière au Vietnam* (CNRS, 2019).

34. Archibald Little, *Through the Yang-Tse Gorges Or Trade and Travel in Western China* (First published 1888; Cambridge University Press, 2010), 7.

35. WWF-UK, "The Yangtze," https://www.wwf.org.uk/where-we-work/places/yangtze. Altogether, by 2022, China had reportedly at least 6,500 large dams and over 95,000 small dams. For India, the second largest dammed nation in the world, the number of completed large dams is estimated at 5,336 and 400 more are under construction. See Yan Gao, *Yangzi Waters: Transforming the Water Regime of the Jianghan Plain in Late Imperial China* (Brill, 2022); Corey Ross, *Liquid Empire* (Princeton University Press, 2024), 358n; Ruth Gamble et al., *Rivers of the Asian Highlands: From Deep Time to the Climate Crisis* (Routledge, 2025), 193–94, 201.

36. S. L. Yang et al., "Decline of Yangtze River Water and Sediment Discharge: Impact from Natural and Anthropogenic Changes," *Scientific Reports* 5, article no. 12581 (2015), https://doi.org/10.1038/srep12581.

37. Henry David Thoreau, *The Writings of Henry David Thoreau*, vol. 8 (Houghton, Mifflin and Company, 1887), 109–10.

38. Astrida Neimanis, *Bodies of Water: Posthuman Feminist Phenomenology* (Bloomsbury, 2017). For a collection of essays on the intimate lifeworld of the river as it negotiates the Anthropocene, see Iftekhar Iqbal, Hasharina Hasan, and Asiyah Kumpoh eds., *Fluid Phenomena: The River and Anthropocene Life-World in the Asia-Pacific* (Brill, forthcoming).

BIBLIOGRAPHY

Archival Sources
The British Library, London
India Office Records (IOR): H, L, M, V Series
European Manuscript: C, E, F Series

National Archives, Kew Gardens, London
Foreign Office Series (FO)
War Office Series (WO)
Prime Minister's Office Series (PREM)

Works published before 1950
Agassiz, A. R. "From Hai-Phong in Tong-King to Canton, Overland." *Proceedings of the Royal Geographical Society and Monthly Record of Geography* 13, no. 5 (May 1891): 249–64.
Ainscough, Thomas M. *Notes from a Frontier*. Kelly and Walsh, 1915.
Alcock, Rutherford. *The Journey of Augustus Raymond Margary, From Shanghae to Bhamo, and Back to Manwyne*. Macmillan and Co., 1876.
Anderson, J. "Irrawaddy and its Sources." Paper read at the Royal Geographical Society, June 13, 1870.
Andree, Karl. *Geographie des Welthandels: Mit geschichtlichen erläuterungen*. Vol. 2. Berlag von Julius Meier, 1872.
Anonymous. "Deutsche Interessen am Yangtze," *Ostasiatischer Lloyd* (August 17, 1900).
Anonymous. "India and China", *The Times*, no. 21327 (January 17, 1853).
Anonymous. "The North-East Frontier of Bengal." *Saturday Review of Politics, Literature, Science and Art* 58, no. 1513 (October 25, 1884).

Anonymous. "Trade Between Burma and Western China." *Amrita Bazar Patrika*, August 28, 1912.
Anonymous. "Western China." *The Nineteenth Century: A Monthly Review* 38 (July–December 1895).
Anonymous. "Western China: Its Products and Trade." *Littell's Living Age* 186, no. 2409 (August 30, 1890).
Baillie, John. *Rivers in the Desert: or, Mission-Scenes in Burmah*. Seeley, Jackson and Halliday, 1859.
Blakiston, Thomas W. *Five Months on the Yang-Tsze*. John Murray, 1862.
Burrard, S. G., and H. H. Hayden. *A Sketch of the Geography and Geology of the Himalayan Mountains and Tibet, part III: The Rivers of the Himalaya and Tibet*. Superintendent Government Press, 1907.
Cadell, Henry M. "A Sail Down the Irrawaddy." *Scottish Geographical Magazine* 15, no. 5 (1901).
Carey, Fred W. "A Trip to the Chinese Shan States." *The Geographical Journal* 14, no. 4 (October 1899).
Chang, Yin-Tang. "Anthropological Features of the Shans and Their Geographical Environment in South-West Yunnan." *Man* 44, no. 55 (May–June 1944).
Christian, John L. "Trans-Burma Trade Routes to China." *Pacific Affairs* 13, no. 2 (1940).
Colquhoun, A. R. *Amongst the Shans*. Field & Tuer, 1885.
Colquhoun, A. R. *Ethnic History of the Shans*. Reprint of *Amongst the Shans*, 1885; Manas Publications, 1985.
Colquhoun, A. R. "Tongquin." *Straits Times Weekly Issue* (October 20, 1883).
Coolidge, Harold Jefferson, and Theodore Roosevelt. *Three Kingdoms of Indo-China*. Thomas Y Crowell Company, 1933.
Cooper, T. T. "Communication Between Assam and China." *Journal of the Society of Arts*, November 4, 1870.
Cooper, T. T. "Letter from Mr. T. T. Cooper, on the Course of the Tsan-Po and Irrawaddy and on Tibet." *Proceedings of the Royal Geographical Society of London* 13, no. 5 (1868–1869).
Cooper, T. T. *The Mishmee Hills. An Account of a Journey Made in an Attempt to Penetrate Thibet from Assam to Open New Routes for Commerce*. Henry S. King & Co. 1873.
Coryton, J. "Trade Routes Between British Burmah and Western China." *Journal of the Royal Geographical Society of London* 45 (1875).
Coryton, J., and Mr. Margary, "Trade Routes Between British Burmah and Western China; with Extracts of Letters from Mr. Margary." *Proceedings of the Royal Geographical Society of London* 19, no. 4 (1874–1875).
Cottam, Henry. "Overland Route to China viâ Assam, Tenga Pani River, Khamti, and Singphoo Country." *Proceedings of the Royal Geographical Society of London* 21, no. 6 (1876–1877).

Cotton, Arthur. "On a Communication Between India and China by the Line of the Burhampooter and Yangtsze." *Journal of the Royal Geographical Society of London* 37 (1867).
Das, Gyanendra Mohon. *Banger Bahire Bangali* [The Bengalis beyond Bengal]. Indian Publishing House, 1931.
Davenport, Arthur. *Report by Mr Davenport Upon the Trading Capabilities of the Country Traversed by the Yunnan Mission.* Harrison and Sons, 1877.
Davies, H. R. *Yün-nan: The Link Between India and Yangtze.* Cambridge University Press, 1909.
Desgodins, C. H. *Mission du Thibet de 1855 à 1870: Relations et observations diverses sur les populations de l'Indo-Chine septentrionale, du Thibet, du pays des Lolos et des autres contrées de la Chine occidentale.* V. Palmé, 1872.
d'Orléans, Prince Henri. *From Tonkin to India by the Sources of the Irawadi, January '95–January '96.* Trans. Hamley Bent. Methuen & Co, 1898.
d'Orléans, Prince Henri. "From Yun-nan to British India." *The Geographical Journal* 7, no. 3 (March 1896).
d'Orléans, Prince Henri. "A Journey from Tonkin by Tali-Fu to Assam." *The Geographical Journal* 8, no. 6 (December 1896): 566–82.
Duckworth, Henry. *New Commercial Route to China (Capt. Sprye's Proposition).* George Philip and Son, 1861.
Dunlop, William B. "The Key of Western China." *Asiatic Quarterly Review* 7 (April 1889).
Dupuis, J. *A Journey to Yunnan and the Opening of the Red River to Trade.* Translated by Walter E. J. Tips. White Lotus Press 1998, originally published in French, 1880.
Edmunds, Charles K. "Some of China's Physical Problems." *The Journal of Race Development* 4, no. 2 (1913): 136–40.
Elias, Ney. "A Visit to the Valley of the Shueli, in Western Yunnan." *Proceedings of the Royal Geographical Society of London* 20, no. 4 (1875–1876): 234–41.
Fitkin, Gretchen Mae. *The Great River: The Story of a Voyage on the Yangtze Kiang.* North-China Daily News & Herald, 1922.
Foreign Office. *Reports from Her Majesty's Consuls on the Manufactures, Commerce &tc.*, vol. 13. Harrison and Sons, 1873.
Furnivall, J. S. *Colonial Policy, and Practice: A Comparative Study of Burma and Netherlands India.* Cambridge University Press, 1948.
Government of Burma. *Gazetteer of Upper Burma and the Shan States*, part 1, vol. 1 Government of Burma, 1900.
Government of Great Britain. *Parliamentary Papers, Great Britain, vol. 49, House of Commons: Accounts and Papers.* Government of Great Britain, 1873.
Government of India. *Imperial Gazetteer of India*, vol. 21. Clarendon Press, 1908.
Government of India. *Imperial Gazetteer: Eastern Bengal and Assam. Mountains, Lakes, Islands, Rivers, Canals, and Historic Areas.* Bengal Secretariat Press, 1906.

Hallett, Holt S. *A Thousand Miles on an Elephant in the Shan States*. William Blackwood and Sons, 1890.

Hallett, Holt S. "Western China," *The Nineteenth Century: A Monthly Review* 38 (July–December 1895).

Hanson, Ola. *The Kachins: Their Customs and Traditions*. First published 1913, Cambridge University Press, 2012.

Hardiman, John Percy. *Gazetteer of Upper Burma and the Shan States*, vol. 1. Government of Burma, 1900.

Hosie, Alexander. *Three Years in Western China: A Narrative of Three Journeys in Ssu-ch'uan, Kuei-chow, and Yün-nan*. George Philip & Son, 1897.

Hunter, W. W. *The Imperial Gazetteer of India*, vol. 8. Trübner and Co., 1886.

Hunter, W. W. *The Imperial Gazetteer of India*, vol. 12. Trübner and Co., 1887.

Hurlbut, G. C. "Geographical Notes." *Journal of the American Geographical Society of New York* 24 (1892): 94–141.

Inspectorate General of Customs. *Decennial Reports on the Trade, Navigation, and Industries, etc., of the Ports Open to Foreign Commerce in China and Corea and on the Condition and Development of the Treaty Port Provinces 1882–91*. Statistical Department of the Inspectorate General of Customs, Shanghai, 1893.

Jamieson, P. E. *Burma Gazetteer: Amherst District*. Vol. A. Government Printing and Stationery, 1913.

Kaulback, Ronald. *Salween*. Hodder and Stoughton, 1938.

Kuo, Tsung-fei. "A Brief History of the Trade Routes Between Burma, Indochina and Yunnan." *T'ien Hsio Monthly* 12, no. 1 (1941).

Lambert, E. T. D. "From the Brahmaputra to the Chindwin." *The Geographical Journal* 89, no. 4 (1937).

Lamington, Lord. "Journey Through the Trans-Salwin Shan States to Tong-King." *Proceedings of the Royal Geographical Society and Monthly Record of Geography* 13, no. 12 (December 1891).

Laurie, W. F. B. "British and Upper Burma, and Western China: Their Concurrent Commercial Interests." *Journal of the Society of Arts* (June 11, 1880).

Little, Archibald. *Across Yunnan: A Journey of Surprises*. Cambridge University Press, 1910.

Little, Archibald. *Across Yunnan and Tonking*. Chungking, 1904.

Little, Archibald. *Through the Yang-tse Gorges, Or Trade and Travel in Western China*. First published 1888. Cambridge University Press, 2010.

Litton, George. *Report on A Journey in North and North-West Yunnan, Season 1902–1903*. Shanghai Mercury, 1903.

Lockhart, William. "On the Importance of Opening the Navigation of the Yang-tse-Kiang, and the Changes That Have Lately Taken Place in the Bed of the Yellow River." *Proceedings of the Royal Geographical 2Society of London* 2, no. 4 (1857–1858): 201–9.

MacMahon, A. R. "Burmese Border Tribes and Trade Routes." *Blackwoods Edinburgh Magazine* 140 (September 1886).

McMahon, A. P. "On Our Prospects of Opening a Route to South-Western China, and Explorations of the French in Tongquin and Cambodia," *Proceedings of the Royal Geographical Society of London* 18, no. 4 (1873–1874)

Manifold, C. C. "The Problem of the Upper Yang-Tze Provinces and Their Communications." *The Geographical Journal* 25, no. 6 (June 1905): 589–617.

Markham. C. R. "Travels in Great Tibet, and Trade Between Tibet and Bengal." *The Journal of the Royal Geographical Society of London* 45 (1875): 299–315.

Martin, Montgomery. *The History, Antiquities, Topography and Statistics, vol. 3: Assam.* W. H. Allen, 1838.

M'Cosh, John. "On the Various Lines of Overland Communication Between India and China." *Proceedings of the Royal Geographical Society of London* 5, no. 2 (1860–1861): 47–55.

McSwiney, J. *Census of India 1911, vol. 3, Assam, Pt. 1, Report.* Assam Secretariate Printing Office, 1912.

Millington, Powell. *On the Track of the Abor.* Smith, Elder & Co., 1912.

Mills, J. P. "The Mishmis of the Lohit Valley, Assam." *The Journal of the Royal Anthropological Institute of Great Britain and Ireland* 82, no. 1 (1952): 1–12.

Mitra, Dinabandhu. *Kamale Kamini.* [কমলে কামিনী]. Nūtana Saṃskṛta Yantra, 1873.

Morrison, George E. *An Australian in China, Being the Narrative of a Quiet Journey Across China to British Burma.* Horace Cox, 1895.

Mouhot, M. Henri. *Travels in the Central Part of the Indo-China*, vol. I. John Murray, 1864.

O'Connor, V. C. Scott. *The Silken East: A Record of Life and Travel in Burma*, vol. 1. Hutchinson & Co. 1904.

Palmer, J. W. *Up and Down the Irrawaddi or The Golden Dagon: Being Passages of Adventure in the Burman Empire.* Rudd & Carleton, 1859.

Parkes, Harry. "Geographical Notes on Siam, with a New Map of the Lower Part of the Menam River." *The Journal of the Royal Geographical Society of London* 26 (1856).

Peal, S. L. "A Peculiarity of the River Names in Assam and Some of the Adjoining Countries." *Journal of the Asiatic Society of Bengal* 48, no. 4 (1879): 261–70.

Percival, William. *The Land of the Dragon: My Boating and Shooting Excursions to the Gorges of Upper Yangtze.* Hurst and Blackett, 1889.

Phayre, A. P. "On the History of the Burmah Race." *Transactions of the Ethnological Society of London* 5 (1867).

Playne, Somerset. *Bengal and Assam Behar and Orissa: Their History, People, Commerce and Industrial Resources.* The Foreign and Colonial Compiling and Publishing Co., 1917.

Pritchard, B. E. A. "A Journey from Myitkyina to Sadiya viâ the N'mai Hka and Hkamti Long." *The Geographical Journal* 43, no. 5 (May 1914): 521–35.

Protherto, G. W., ed. *Tibet.* H.M. Stationery Office, 1920.

Reh, Khu Oo. "Karenni People at a Glance." In *Citizenship in Myanmar: Ways of Being in and from Burma*, edited by Ashley South and Marie Lall. Cambridge University Press, 2017.

Rennell, James. "An Account of the Ganges and the Burrampooter Rivers." *Philosophical Transactions of the Royal Society of London* 71 (1781): 87–114.

Richthofen, F. v. "Recent Attempts to Find a Direct Trade-Road to South-Western China." In *Ocean Highways: The Geographical Record*, edited by C. R. Markham. N. Trubner and Company, 1874.

Robison, John. *A System of Mechanical Philosophy*, vol. 2, John Murray, 1822.

Sarel, Henry Andrew. "Notes on the Yang-tsze-Kiang, from Han-kow to Ping-shan." *The Journal of the Royal Geographical Society of London* 32 (1862): 1–25.

Scone, G. Colquhoun. *Journal of Lieut. G. Colquhoun Scone, While Employed on the Salween Surveying Expedition*. Military Orphan Press, 1865.

Scott, J. George, ed. *Gazetteer of Upper Burma and the Shan States*, part. 1, vol. 1. Government of British Burma, 1900.

Scott, J. George. "The Hill-Slopes of Tong-Kin." *Proceedings of the Royal Geographical Society and Monthly Record of Geography* 8, no. 4 (April 1886): 217–45.

Shakespear, L. W. *History of Upper Assam, Upper Burmah and North-Eastern Frontier*. Macmillan and Co., 1914.

Simmonds, Peter Lund, and William Henry Giles Kingston, eds. *Colonial Magazine and East India Review* 3 (September–December 1844)

Sladen, E. B. "Expedition from, Burma, via the Irrawaddy and Bhamo, to South-Western China." *The Journal of the Royal Geographical Society of London* 41 (1871).

Sladen, E. B. *Official Narrative of and Papers Connected with the Expedition to Explore the Trade Routes to China via Bhamo*. British Burma Press, 1879.

Sladen, E. B., and Horace Browne. *Mandalay to Momien: A Narrative of the Two Expeditions to Western China of 1868 and 1875*. Macmillan and Co., 1876.

Sprye, R. H. F. "Communication with the South-West Provinces of China from Rangoon in British Pegue." *Proceedings of the Royal Geographical Society of London*, Session 1860–61.

Stamp, L. Dudley. "The Irrawaddy River," *The Geographical Journal* 95, no. 5 (May 1940).

Swinhoe, R. "The Yang-Tsze-Kiang from Tung-Ting Lake to Chung-King to Accompany the Paper by R. Swinhoe," *Journal of the Royal Geographical Society of London* 40 (1870).

Thoreau, Henry David. *The Writings of Henry David Thoreau*, vol. 8. Houghton, Mifflin and Company, 1887.

Thorp, Ellen. *Quiet Skies on Salween*. Jonathan Cape, 1945.

Tsung-fei, Kuo. "A Brief History of the Trade Routes Between Burma, Indochina and Yunnan." *T'ien Hsio Monthly* 12 (1941).

Ward, F. Kingdon. *In Farthest Burma*. Seeley, Service & Co., 1921.

Ward, F. Kingdon. *The Land of the Blue Poppy*. Cambridge University Press, 1913.

Ward, F. Kingdon. "The Seinghku and Delei Valleys, North-East Frontier of India." *The Geographical Journal* 75, no. 5 (May 1930): 412–32.

Ward, F. Kingdon. "The Valleys of the Kham." *The Geographical Journal* 56, no. 3 (September 1920).

Ward, F. Kingdon, and Malcolm Smith, "The Himalaya East of the Tsangpo." *The Geographical Journal* 84, no. 5 (November 1934): 369–94.
Williamson, Noël. "The Lohit-Brahmaputra Between Assam and South-Eastern Tibet, November, 1907, to January, 1908." *The Geographical Journal* 34, no. 4 (1909).
Wilson, E. H. *A Naturalist in Western China*, vol. 1. Methuen & Company, 1913.
Wingate, A. M. S. "Recent Journey from Shanghai to Bhamo through Hunan," *The Geographical Journal* 14, no. 6 (December 1899).
Woodthorpe, R. G. "The Country of the Shans." *The Geographical Journal* 7, no. 6 (June 1896).
Woodthorpe, R. G. "Explorations on the Chindwin River, Upper Burma." *Proceedings of the Royal Geographical Society and Monthly Record of Geography*, New Monthly Series 11, no. 4 (April 1889): 197–216.
Younghusband, G. J. *Eighteen Hundred Miles on A Burmese Tat: Through Burmah, Siam, and the Eastern Shan States*. W. H. Allen and Co., 1888; reprint Asian Educational Services, 1995.
Yule, Henry. "On the Geography of Burma and Its Tributary States, in Illustration of a New Map of Those Regions." *The Journal of the Royal Geographical Society of London* 27 (1857): 54–108.

Works published Since 1950

Adas, Michael. *Burma Delta: Economic Development and Social Change on an Asian Rice Frontier 1852–1941*. University of Wisconsin Press, 1974.
Aiyadurai, A. "The Meyor: A Least Studied Frontier Tribe of Arunachal Pradesh, Northeast India." *The Eastern Anthropologist* 64, no. 4 (2011): 459–503.
Aiyadurai, Ambika, and Claire Seungeun Lee. "Living on the Sino-Indian Border: The Story of the Mishmis in Arunachal Pradesh, Northeast India." *Asian Ethnology* 76, no. 2 (2017): 367–95.
Aiyar, Sana. "Revolutionaries, Maulvis, and Monks: Burma's Khilafat Moment." In *Oceanic Islam*, edited by Sugata Bose and Ayesha Jalal. Bloomsbury, 2020.
Alden, Chris, and Elizabeth Sidiropolos. "Silk, Cinnamon and Cotton: Emerging Power Strategies for the Indian Ocean and the Implications." South African Institute of International Affairs, *Policy Insights* 18 (June 2015). https://saiia.org.za/wp-content/uploads/2015/06/Policy-Insights-18.pdf. Accessed February 6, 2025.
Amin, Ash, and Philip Howell. "Thinking the Commons." In *Releasing the Commons: Rethinking the Futures of the Commons*, edited by Ash Amin and Philip Howell. Routledge, 2016.
Amrith, Sunil S. *Crossing the Bay of Bengal: The Furies of Nature and the Fortunes of Migrants*. Harvard University Press, 2013.
Amrith, Sunil S. *Unruly Waters: How Rains, Rivers, Coasts, and Seas Have Shaped Asia's History*. Basic Books, 2018.
Amrith, Sunil, and Dan Smyr Yü. "The Himalaya and Monsoon Asia: Anthropocenic Climes Since the 1800s." In *Storying Multipolar Climes of the Himalaya, Andes and Arctic: Anthropocenic Climate and Shapeshifting Watery Lifeworlds*, edited by Dan Smyer Yü and Jelle J. P. Wouters. Routledge, 2023.

Atwill, David G. *The Chinese Sultanate: Islam, Ethnicity, and the Panthay Rebellion in Southwest China, 1856–1873.* Stanford University Press, 2006.
Baruah, Sanjib. *In the Name of the Nation.* Stanford University Press, 2020.
Bejabaruwā, Lakshmīnātha. *jivan smoron.* Assam Book Trust, 1992.
Bello, David. *Across Forest Steppe, and Mountain: Environment, Identity, and Empire in Qing China's Borderlands.* Cambridge University Press, 2016.
Benedict, Carol. *Bubonic Plague in Nineteenth-Century China.* Stanford University Press, 1996.
Berlie, Jean. "Links Between Muslims in Yunnan and Northern Thailand." In *Where China Meets Southeast Asia: Social and Cultural Change in the Border Regions,* edited by Grant Evans, Christopher Hutton, and Kuah Khun Eng. Institute of Southeast Asian Studies, 2000.
Bhattacharya, Avijeet. *Journeys on the Silk Road Through Ages—Romance, Legend, Reality.* Zorba Books, 2017.
Biggs, David. *Quagmire: Nation-Building and Nature in the Mekong Delta.* Washington University Press, 2020.
Bird-David, Nurit. "Animism' Revisited: Personhood, Environment, and Relational Epistemology." *Current Anthropology* 40 (Supplement, February 1999).
Blackbourn, David. "Time Is a Violent Torrent: Constructing and Reconstructing Rivers in Modern German History," In *Rivers in History: Perspectives in Europe and North America,* edited by Christof Mauch and Thomas Zeller. University of Pittsburgh Press, 2008.
Blue, Archibald Duncan. "The China Coast: A Study of British Shipping in Chinese Waters 1842–1914." PhD diss., University of Strathclyde, 1982.
Blue, A. D. "Land and River Routes to West China (With Especial Reference to the Upper Yangtze)," *Journal of the Hong Kong Branch of the Royal Asiatic Society* 16 (1976): 162–78.
Bollier, David, and Silke Helfrich, eds. *Patterns of Commoning.* The Commons Strategies Group, 2015.
Bollier, David, and Silke Helfrich, eds. *The Wealth of the Commons: A World Beyond Market and State.* The Commons Strategy Group, 2012.
Bose, Sugata. *A Hundred Horizons: Indian Ocean in the Age of Global Empire.* Harvard University Press, 2009.
Braudel, Fernand. *The Mediterranean and the Mediterranean World in the Age of Philip II,* vol. 1. University of California Press, 1995.
Bray, John. "Trade, Territory, and Missionary Connections in the Sino-Tibetan Borderlands." In *Frontier Tibet: Patterns of Change in the Sino-Tibetan Borderlands,* edited by Stéphane Gros. Amsterdam University Press, 2019.
Brocheux, Pierre, and Daniel Hémery. *Indochina: An Ambiguous Colonization, 1858–1954.* University of California Press, 2011.
Bronson, Bennet. "Exchange at the Upstream and Downstream Ends: Notes Toward a Functional Model of the Coastal State in Southeast Asia." In *Economic Exchange and Social Interaction in Southeast Asia: Perspectives from Prehistory, History, and*

Ethnography, edited by Karl Hutterer. University of Michigan, Center for South and Southeast Asian Studies, 1978.
Brown, Ian. *A Colonial Economy in Crisis: Burma's Rice Cultivators and the World Depression of the 1930s*. Routledge, 2005.
Buckley, Michael. *Meltdown in Tibet*. Palgrave Macmillan, 2014.
Burling, Robbins. *Hill Farms and Padi Fields: Life in Mainland Southeast Asia*. Englewood Prentice-Hall, 1965.
Carr, E. Summerson, and Michael Lempert, eds. *Scale: Discourse and Dimensions of Social Life*. University of Chicago Press, 2016.
Cederlöf, Gunnel, *Founding an Empire on India's North-Eastern Frontiers, 1790–1840: Climate, Commerce, Polity*. Oxford University Press, 2013.
Chabrowski, Igor Iwo. "'Tied to a Boat by the Sound of a Gong": World, Work and Society Seen through the Work Songs of Sichuan Boatmen (1880–1930s)." PhD diss., European University Institute, 2013.
Chandran, J. "The Burma-Yunnan Railway: Anglo-French Rivalry in Mainland Southeast Asia and South China, 1895–1902." Ohio University Center for International Studies, 1971.
Chang, Wen-Chin. *Beyond Borders: Stories of Yunannese Chinese Migrants of Burma*. Cornell University Press, 2014.
Chang, Wen-Chin. "Circulations via Tangyan, a Town in the Northern Shan State of Burma." In *Asia Inside Out: Connected Places*, edited by Eric Tagliacozzo, Helen F. Siu, and Peter Perdue. Harvard University Press, 2015.
Chang, Wen-Chin. "From a Shiji Episode to the Forbidden Jade Trade During the Socialist Regime in Burma." In *Chinese Circulations: Capital, Commodities, and Networks in Southeast Asia*, edited by Eric Tagliacozzo and Wen-Chin Chang. Duke University Press, 2011.
Chaudhuri, KN. *Trade and Civilizations in the Indian Ocean: An Economic History*. Cambridge University Press, 1985.
Chellaney, Brahma. *Water: Asia's New Battleground*. Georgetown University Press, 2011.
Chen, Fei. "Rediscovering the Yellow River and the Yangtze River: The Circulation of Discourses on the North-South Dichotomy Between Late Qing China and Meiji Japan." *International Journal of Asian Studies* 16, no. 1 (2019): 33–51.
Chi, Ch'ao-ting. *Wartime Economic Development of China*. Garland,1980.
Christian, John L. "Trans-Burma Trade Routes to China." *Pacific Affairs* 13, no. 2 (June 1940): 173– 219.
Christie, Jan Wissenman. "Water and Rice in Early Java and Bali." In *A World of Water: Rain, Rivers, and Seas in Southeast Asian Histories*, edited by Peter Boomgaard. KITLV, 2007.
Clarence-Smith, William G. "Breeding and Power in Southeast Asia Horses, Mules and Donkeys in the Longue Durée." In *Environment, Trade and Society in Southeast Asia*, edited by David Henley and Henk Schulte Nordholt. Brill, 2015.

Clark, M. K., L. M. Shoenbohm, L. H. Royden, et al. "Surface Uplift, Tectonics, and Erosion of Eastern Tibet from Large-Scale Drainage Patterns." *Tectonics* 23, no. 4 (2004).
Cooke, Nola, and Li Tana, eds. *Water Frontier: Commerce and the Chinese in the Lower Mekong Region, 1750–1880* (University of Hawai'i Press, 2004).
Cooke, Nola, Li Tana, and James A. Anderson, eds. *The Tongking Gulf Through History*. University of Pennsylvania Press, 2011.
Culas, Christian. "Migrants, Runaways and Opium Growers: Origins of the Hmong in Laos and Siam in the Nineteenth and Early Twentieth Centuries." In *Turbulent Times and EnDuring Peoples: Mountain Minorities in the South-East Asian Massif*, edited by Jean Michaud and Jan Ovesen. Curzon, 2000.
Cunha, Dilip da. *The Invention of Rivers: Alexander's Eye and Ganga's Descent*. University of Pennsylvania Press, 2019.
Daniels, Christian, and Jianxiong Ma. "Introduction: The Agency of Local Elites in the Transformation of Western Yunnan During the Ming Dynasty." In *The Transformation of Yunnan in Ming China: From the Dali Kingdom to Imperial Province*, edited by Christian Daniels and Jianxiong Ma. Routledge, 2020.
Dao, Nga, and Anh Tuan Le, eds. *Vietnam Hydropower and Its Challenges to Sustainability*. Science and Technology Publisher, 2016.
Davis, Bradley Camp. *Imperial Bandits: Outlaws and Rebels in the China-Vietnam Borderlands*. University of Washington Press, 2017.
Dean, Karin. *The Kachin Tackling the Territorial Trap: A Nation Divided By the Sino-Myanmar Boundary*. VDM, 2010.
Derr, Jennifer L. *The Lived Nile: Environment, Disease, and Material Colonial Economy in Egypt*. Stanford University Press, 2019.
Di, Wang. *Kuachu fengbi de shijie: Changjiang shangyou quyu shehui yanjiu, 1644–1911* [Striking out of a closed world: A study on society in the upper Yangzi region, 1644–1911]. Zhonghua Shuju, 1993.
Dick, Howard, and Peter J. Rimmer. *Cities, Transport and Communications: The Integration of Southeast Asia Since 1850*. Palgrave Macmillan, 2003.
Dijk, Kees van. *Pacific Strife*. University of Amsterdam, 2015.
Duara, Prasenjit. *The Crisis of Global Modernity*. Cambridge University Press, 2014.
Durrenberger, E. Paul. "Lisu: Political Form, Ideology and Economic Action." In *Highlanders of Thailand*, edited John McKinnon and Wanat Bhruksasri. Oxford University Press, 1986.
Eberspächer, Cord. *Die deutsche Yangtse-Patrouille. Deutsche Kanonenbootpolitik in China im Zeitalter des Imperialismus 1900–1914*. Verlag Dr. Dieter Winkler, 2004.
Eyler, Brian. *Last Days of the Mighty Mekong*. Zed Books, 2019.
Failler, Philippe Le. "The Dêo Family of Lai Châu: Traditional Power and Unconventional Practices." *Journal of Vietnamese Studies* 6, no. 2 (2011).
Failler, Philippe Le. *La rivière Noire: L'intégration d'une marche frontière au Vietnam*. CNRS, 2019.

Fleming, Sean W. *Where the River Flows: Scientific Reflections on Earth's Waterways.* Princeton University Press, 2017.
Forbes, Andrew D. W. "The "ČĪN-HỌ̆" (Yunnanese Chinese) Caravan Trade with North Thailand During the Late Nineteenth and Early Twentieth Centuries." *Journal of Asian History* 21, no. 1 (1987): 1–47.
Forbes, Andrew, and David Henley. *The Haw: Traders of the Golden Triangle.* Asia Film House, 1997.
Forbes Andrew, and David Henley. *Traders of Golden Triangle.* Cognoscenti Books, 2013.
Gaenszle, Martin. "Where the Waters Dry Up—The Place of Origin in Rai Myth and Ritual." In *Origins and Migrations in the Extended Eastern Himalayas*, edited by Toni Huber and Stuart Blackburn. Brill, 2012.
Gagoi, Dilip. *Making of India's Northeast: Geopolitics of Borderland and Transnational Interactions.* Routledge, 2019.
Gamble, Ruth, Gillian G. Tan, Hongzhang Xu, et al. *Rivers of the Asian Highlands: From Deep Time to the Climate Crisis.* Routledge, 2025.
Gao, Yang. *Yangzi Waters: Transforming the Water Regime of the Jianghan Plain in Late Imperial China.* Brill, 2022.
Gardner, Kyle J. *The Frontier Complex: Geopolitics and the Making of the India-China Border.* Cambridge University Press, 2021.
Geusau, Leo Alting Von. "Dialectics of Akhazang: The Interiorizations of a Perennial Minority Group." In *Highlanders of Thailand*, edited by John McKinnon and Wanat Bhruksasri. Oxford University Press, 1986.
Ghose, Amitav. *The Glass Palace.* Random House, 2000.
Ghose, Devleena. "Burma-Bengal Crossings: Intercolonial Connections in Pre-Independence India." *Asian Studies Review* 40, no. 2 (2016): 156–72.
Giersch, C. Patterson. "Across 'Zomia' with Merchants, Monks, and Musk: Process Geographies, Trade Networks, and the Inner-East-Southeast Asian Borderlands." *Journal of Global History* 5, no. 2 (2010): 215–39.
Giersch, C. Patterson. *Asian Borderlands: The Transformation of Qing China's Yunnan Frontier.* Harvard University Press, 2006.
Gilmartin, David. *Blood and Water: The Indus River Basin in Modern History.* University of California Press, 2015.
Government of Burma. *Burma Gazetteer, Salween District*, vol. A., Government Printing and Stationery Press, Union of Burma, 1957.
Gungwu, Wang. "A Two-Ocean Mediterranean." In *Anthony Reid and the Study of the Southeast Asian Past*, edited by Geof Wade and Tana Li. ISEAS-Yusof Ishak Institute, 2012.
Gupta, Harish, Shu-Ji Kao, and Minhan Dai. "The Role of Mega Dams in Reducing Sediment Fluxes: A Case Study of Large Asian Rivers." *Journal of Hydrology* 464–65 (September 25, 2012): 447–58.
Gupta, Swarupa. *Cultural Constellations, Place-Making and Ethnicity in Eastern India, c. 1850–1927.* Brill, 2017.

Guyot-Réchard, Berenice. *Shadow States. India, China and the Himalayas, 1910–1962*. Cambridge University Press, 2017.
Hamid, Mohamed Effendy Bin Abdul. "Understanding the Cham Identity in Mainland Southeast Asia: Contending Views." *SOJOURN: Journal of Social Issues in Southeast Asia* 21, no. 2 (2006): 230–53.
Haraway, Donna J. *Staying with the Trouble: Making Kin in the Chthulucene*. Duke University Press, 2016.
Hardin, Garrett. "Tragedy of the Commons." *Science* 162, no. 3859 (1968): 1243–48.
Heidegger, Martin. *Poetry, Language, Thought*. Translated by Albert Hofstadter. Harper Colophon Books, 1971.
Hengsuwan, Paiboon. "Explosive Border: Dwelling, Fear and Violence on the Thai-Burmese Border Along the Salween River." *Asia-Pacific Viewpoint* 54, no. 1 (2013).
Hevia, James L. *Animal Labour and Colonial Warfare*. University of Chicago Press, 2018.
High, Holly. "Dreaming Beyond Borders: The Thai/Lao Borderlands and the Mobility of the Marginal." In *On the Borders of State Power: Frontiers in the Greater Mekong Sub-Region*, edited by Martin Gainsborough. Routledge, 2009.
Hill, Ann Maxwell. Merchants and Migrants: Ethnicity and Trade Among Yunnanese Chinese in Southeast Asia. Yale University Press, 1998.
Hill, Ann Maxwell. "The Yunnanese: Overland Chinese in Northern Thailand." In *Highlanders of Thailand*, edited by John McKinnon and Wanat Bhruksasri. Oxford University Press, 1986.
Hla, Nai Pan. "The Significant Role of the Mon Language and Culture in Southeast Asia." Part 1. Institute for the Study of Languages and Cultures of Asia and Africa (ILCAA), 1992.
Hlaing, U. Chit. "Anthropological Communities of Interpretation for Burma: An Overview." *Journal of Southeast Asian Studies* 39, no. 2 (June 2008): 239–54.
Ho, Selina., David M. Lampton, and Cheng-Chwee Kuik. *Rivers of Iron: Railroads and Chinese Power in Southeast Asia*. University of California Press, 2020.
Htoo, Aung Kyaw. "Myanmar Petroleum Sector Future Pathways and Prospect," Ministry of Energy, 2014. https://web.archive.org/web/20150417055259/http://www.myanmar-oilgas.com/OilGas/media/Site_Images/Day-2-AM-1-EPD-Petroleum-Sector-Prospects.pdf.
Huber, Toni. "Micro-Migrations of Hill Peoples in Northern Arunachal Pradesh: Rethinking Methodologies and Claims of Origins in Tibet." In *Origins and Migrations in the Extended Eastern Himalayas*, edited by Toni Huber and Stuart Blackburn. Brill, 2012.
Hudson, Paul F. *Flooding and Management of Large Fluvial Lowlands: A Global Environmental Perspective*. Cambridge University Press, 2021.
Iqbal, Iftekhar. *The Bengal Delta: Ecology, State and Social Change 1840–1943*. Palgrave, 2010.

Iqbal, Iftekhar. "The Bengali Muslim: Language and Space-Making at the Ocean's Margins." In *Oceanic Islam*, edited by Sugata Bose and Ayesha Jalal. Bloomsbury, 2020.

Iqbal, Iftekhar. "The Boat Denial Policy and the Great Bengal Famine." *Journal of the Asiatic Society of Bangladesh* 56, no. 1–2 (2011): 271–82.

Iqbal, Iftekhar. "From Zomia to Holon: Rivers and Transregional Flows in Mainland Southeastern Asia, 1840–1950." *Suvannabhumi: Multidisciplinary Journal of Southeast Asian Studies* 12, no. 2 (2020).

Iqbal, Iftekhar. "Locating the Riparian Commons in Eastern South Asia: A Translocal Perspective." In *Urban Development and Environmental History in Modern South Asia*, edited by Ian Talbot and Amit Ranjan. Routledge, 2022.

Iqbal, Iftekhar. "Reclaiming the Crossroads Between India and China: A View from the River." *Economic and Political Weekly* 49, no. 51 (2009).

Iqbal, Iftekhar. "The Space Between Nation and Empire: Making and Unmaking of Eastern Bengal and Assam, 1905–1911." *Journal of Asian Studies* 74, no. 2 (2015).

Iqbal, Iftekhar, Hasharina Hasan, and Asiyah Kumpoh eds., *Fluid Phenomena: The River and Anthropocene Life-World in the Asia-Pacific* (Brill, forthcoming).

Izikowitz, Karl Gustav. *Lamet: Hill Peasants in French IndoChina*. AMS, 1979.

Johnson, Andrew Alan. *Mekong Dreaming: Life and Death Along a Changing River*. Duke University Press, 2020.

Kacha-Ananda, Chob. "Migration, Settlements and Land." In *Highlanders of Thailand*, edited by John McKinnon and Wanat Bhruksasri. Oxford University Press, 1986.

Kar, Bodhisattva. "Nomadic Capital and Speculative Tribes: A Culture of Contracts in the Northeastern Frontier of British India." *The Indian Economic and Social History Review* 53, no. 1 (2016): 41–67.

Karlsson, Bengt G. "Evading the State: Ethnicity in Northeast India Through the Lens of James Scott." *Asian Ethnology* 72, no. 2 (2013).

Kaske, Elisabeth. "Sichuan as a Pivot: Provincial Politics and Gentry Power in Late Qing Railway Projects in Southwestern China." In *Southwest China in a Regional and Global Perspective (c.1600–1911)*, edited by Ulrich Theobald and Cao Jin. Brill, 2018.

Khan, Naveeda. "The River and the Corruption of Memory." *Contributions to Indian Sociology* 49, no. 3 (2015): 389–409.

Kim, Nanny. "Fuel for the Smelters: Copper Mining and Deforestation in Northeastern Yunnan During the High Qing, 1700 to 1850." In *Southwest China in Regional and Global Perspectives (c. 1600–1911)*, edited by Ulrich Theobold and Cao Jin. Brill, 2018.

Kim, Nanny. *Mountain Rivers, Mountain Roads: Transport in Southwest China, 1700–1850*. Brill, 2020.

Külz, Helmut R. "Further Water Disputes Between India and Pakistan." *The International and Comparative Law Quarterly* 18, no. 3 (July 1969): 718–38.

Laffey, Ella S. "French Adventures and Chinese Bandits in Tonkin: The Garnier Affair in Its Local Context." *Journal of Southeast Asian Studies* 6, no. 1 (March 1975): 38–51.

Lange, Diana. *An Atlas of the Himalayas by a 19th Century Tibetan Lama: A Journey of Discovery*. Brill, 2020.
Lange, Diana. "'The Boatman Is More Beautiful than a God': Poetizing and Singing on the Rivers in Central and Southern Tibet." In *The Illuminating Mirror: Tibetan Studies in Honour of Per K. Soerensen on the Occasion of His 65th Birthday*, edited by Guntram Hazod and Olaf Czaja. Ludwig Reichert Verlag, 2016.
Latour, Bruno. "On Actor-Network Theory: A Few Clarifications." *Soziale Welt* 47, no. 4 (1996): 369–81.
Leach, Edmund. *Political System of Highland Burma*. Oxford University Press, 1968.
Lentz, Christian C. *Contested Territory: Điện Biên Phủ and the Making of Northwest Vietnam*. Yale University Press, 2019.
Leonard, John, "Books of the Times," *The New York Times*, May 20, 1982.
Li, Tana. "Between Mountains and the Sea: Trades in Early Nineteenth-Century Northern Vietnam." *Journal of Vietnamese Studies* 7, no. 2 (2012).
Li, Tana. "A Historical Sketch of the Landscape of the Red River Delta." *TRaNS: Trans-Regional and -National Studies of Southeast Asia* 4, no. 2 (July 2016): 352–60.
Li, Tana. "Towards an Environmental History of the Eastern Red River Delta, Vietnam, c. 900–1400." *Journal of Southeast Asian Studies* 45, no. 3 (October 2014).
Li, Yi. "Transformation of the Yunnanese Community Along the Sino-Burma Border During the Nineteenth and Early Twentieth Centuries." In *Imperial China and its Southern Neighbours*, edited by Victor H. Mair and Liam Kelley. ISEAS-Yusof Ishak Institute, 2018.
Lin, Chang-Kuan. "Chinese Muslims of Yunnan, Southwest China, With Special Reference to Their Revolt 1885–1873." PhD diss., University of Aberdeen, 1991.
Ludden, David. "India's Spatial History in the Brahmaputra-Meghna River Basin." In *Landscape, Culture, and Belonging: Writing the History of Northeast*, edited by Neeladri Bhattacharya and Joy L. K. Pachuau. Cambridge University Press, 2019.
Ludden, David. "Investing in nature around Sylhet." *Economic and Political Weekly* 38, no. 48 (2003).
Ludden, David. "The Process of Empire: Frontiers and Borderlands." In *Tributary Empires in Global History*, edited by Peter Fibiger Bang and C. A. Bayly. Springer, 2020.
Luong, Hy Van. "Mobile Trading Network from Central Coastal Vietnam." In *Traders in Motion: Identities and Contestations in the Vietnamese Marketplace*, edited by Kirsten W. Endres and Ann Marie Leshkowich. Cornell University Press. 2018.
Lyngdoh, R. S. *John Robert's Ka Histori Ka Thoh Ka Thar*. H. W. Sten, 1979.
Ma, Jianxiong. "The Zhaozhou *Bazi* Society in Yunnan: Historical Process in the *Bazi* Basin Environmental System During the Ming Period (1368–1643)." In *Environmental History in East Asia: Interdisciplinary Perspectives*, edited by Tsui-jung Liu. Routledge, 2014.
Ma, Jianxiong, and Cunzhao Ma. "The Mule Caravans of Western Yunnan: An Oral History of the Muleteers of Zhaozhou." *Transfers* 4, no. (2014): 24–42.
Ma, Jianxiong, and Cunzhao Ma. "The Mule Caravans as Cross-Border Networks: Local Bands and Their Stretch on the Frontier Between Yunnan and Burma." In

*Myanmar's Mountain and Maritime Borderscapes: Local Practices, Boundary-Making and Figured World*s, edited by Su-Ann Ho. ISEAS-Yusof Ishak Institute, 2016.

Mǎ, Lìqín. *Chángjiāng huánghé* [Yangtze River and Yellow River]. National Publication Foundation, 2014.

Mann, Michael. *British Rule on Indian Soil: North India in the First Half of the Nineteenth Century.* Manohar, 1999.

Mauch, Christof, and Thomas G. Zeller, eds. *Rivers in History: Perspectives in Europe and North America.* University of Pittsburgh Press, 2008.

McCrae, Alister, and Alan Prentice. *Irrawaddy Flotilla.* James Paton, 1978.

McElwee, Pamela. *Forests Are Gold: Trees, People, and Environmental Rule in Vietnam.* University of Washington Press, 2016.

McNeill, J. R. *The Mountains of the Mediterranean World: An Environmental History.* Cambridge University Press, 1992.

McGrath,Thomas E. "A Warlord Frontier: The Yunnan-Burma Border Dispute, 1910–1937." Organization of American Historians (OAH) Proceedings, 2003.

Meyer-Fong, Tobie. *What Remains: Coming to Terms with Civil War in 19th Century China.* Stanford University Press, 2013.

Michaud, Jean. "French Military Ethnography in Colonial Upper Tonkin (Northern Vietnam), 1897–1904." *Journal of Vietnamese Studies* 8, no. 4 (Fall 2013): 1–46.

Michaud, Jean. "French Missionary Expansion in Colonial Upper Tonkin," *Journal of Southeast Asian Studies* 35, no. 2 (June 2004): 301–3.

Michaud, Jean. *'Incidental' Ethnographers: French Catholic Missions on the Tonkin-Yunnan Frontier, 1880s–1930.* Brill 2007.

Michaud, Jean. "The Montagnards and the State in Northern Vietnam from 1802–1975. A Historical Overview." *Ethnohistory* 47, no. 2 (2000).

Michaud, Jean, ed. *Turbulent Times and Enduring Peoples: Mountain Minorities in the South-East Asian Massif.* Routledge, 2000.

Michaud, Jean. "Zomia and Beyond." In *Routledge Handbook of Asian Borderlands*, edited by A. Horstmann, M. Saxer, and A. Rippa. Routledge, 2018.

Michaud, Jean, and Christian Culas. "The Hmong of the Southeast Asia Massif: Their Recent History of Migration." In *Where China Meets Southeast Asia: Social and Cultural Change in the Border Regions,* edited by Grant Evans, Christopher Hutton, and Kuah Khun Eng. Institute of Southeast Asian Studies, 2000.

Michaud, Jean, and Sarah Turner, "Tonkin's Uplands at the Turn of the 20th Century: Colonial Military Enclosure and Local Livelihood Effects." *Asia-Pacific View Point* 57, no. 2 (2016).

Middleton, Carl, and Vanessa Lamb, eds. *Knowing the Salween River: Resource Politics of a Contested Transboundary River.* Springer Open, 2019.

Miller, Michelle Ann, Carl Middleton, Jonathan Rigg, and David Taylor, "Hybrid Governance of Transboundary Commons: Insights from Southeast Asia." *Annals of the American Association of Geographers* 110, no. 1 (2020): 297–313.

Mikhail, Alan. *The Animal in Ottoman Egypt.* Oxford University Press, 2014.

Misra, Sanghamitra. "Spaces, Borders, Histories: Identity Construction in Goalpara." PhD diss., University of London, 2004.

Molle, Francois. *Contested Waterscapes in the Mekong Region: Hydropower, Livelihoods and Governance.* Routledge, 2012.

Mollon, J. D. "The Origins of the Concept of Interference." *Philosophical Transactions* 360, no. 1790 (2002).

Mon, Pon Nya. "Ethnic Identity and Political Autonomy of the Mon." In *The Mon Over Two Millennia: Monuments, Manuscripts, Movements,* edited by Patrick McCormick, Mathias Jenny, and Chris Baker. Institute of Asian Studies, 2011.

Morrison, Toni. "The Site of Memory." In *Inventing the Truth: The Art and Craft of Memoir,* edited by William Zinsser. Houghton Mifflin, 1995.

Mostern, Ruth. *The Yellow River: Natural and Unnatural History.* Yale University Press, 2021.

Mukherjee, Rila, ed. *Pelagic Passageways: The Northern Bay of Bengal Before Colonialism.* Primus Books, 2011.

Myint-U, Thant. *Where China Meets India: Burma and the New Crossroads of Asia.* Faber and Faber, 2012.

Nayak, Prateep Kumar, ed. *Making Commons Dynamic: Understanding Change Through Commonisation and Decommonisation.* Routledge, 2021.

Neimanis, Astrida. *Bodies of Water: Posthuman Feminist Phenomenology.* Bloomsbury, 2017.

Osborne, Milton. *Mekong: Turbulent Past, Uncertain Future.* Allen & Unwin, 2006.

Osborne, Milton. *River Road to China. The Mekong River Expedition 1866–1873.* Readers Union, 1976.

Ostrom, Elinor. *Governing the Commons: The Evolution of Institutions for Collective Action.* Cambridge University Press, 1990.

Pan, Yiying. "Nexus of Self-Organization: The Expansion of Collective Responsibility Networks Among Boatmen in Nineteenth-Century Chongqing," *International Journal of Asian Studies* 20, no. 1 (2023).

Pachuau, Joy. *Entangled Lives: Human-Animal-Plant Histories of the Eastern Himalayan Triangle.* Cambridge University Press, 2022.

Pemberton, R. Boileau. "Abstract of the Journal of a Route Travelled by Captain S. F. Hannay, of the 40th Regiment, Native Infantry, in 1835–36, from the Capital of Ava to the Amber Mines of the Hukong Valley on the South-East Frontier of Assam." *SOAS Bulletin of Burma Research* 3, no. 1 (Spring 2005).

Perdue, Peter C. "Crossing Borders in Imperial China." In *Asia Inside Out: Connected Places,* edited by Eric Tagliacozzo, Helen F. Siu, and Peter C. Perdue. Harvard University Press, 2015.

Peterson, Maya K. *Pipe Dreams: Water and Empire in Central Asia's Aral Sea Basin.* Cambridge University Press, 2019.

Pietz, David A. *The Yellow River: The Problem of Water in Modern China.* Harvard University Press, 2015.

Platt, Stephen. *Autumn in the Heavenly Kingdom: China, the West, and the Epic Story of the Taiping Civil War*. Alfred A. Knopf, 2012.
Pomeranz, Kenneth. "The Great Himalayan Watershed: Water Shortages, Mega-Projects and Environmental Politics in China, India, and Southeast Asia." *The Asia-Pacific Journal* 7, no. 30 (July 2009).
Pommaret, Françoise. "Ancient Trade Partners: Bhutan, Cooch Bihar, and Assam (17th–19th centuries)." *Journal of Bhutan Studies* 2, no. 1 (Autumn 2000). http://www.dspace.cam.ac.uk/handle/1810/227005. Accessed February 5, 2025.
Porcher, Jocelyne. "Animal Work." In *A Cultural History of Animals. Vol. 5: In the Age of Empire*, edited by Kathleen Kete. Bloomsbury, 2011.
Reh, Khu Oo. "Karenni People at a Glance." In *Citizenship in Myanmar: Ways of Being in and from Burma*, edited by Ashley South and Marie Lall. Cambridge University Press, 2017.
Reid, Anthony. *Southeast Asia in the Age of Commerce*. Vol. 1. Yale University Press, 1988.
Reid, Anthony. *Southeast Asia in the Age of Commerce*. Vol. 2. Yale University Press, 1993.
Reinhardt, Anne. *Navigating Semi-Colonialism: Shipping, Sovereignty, and Nation-Building in China, 1860–1937*. Harvard University Press 2018.
Reinhardt, Anne. "Treaty Ports as Shipping Infrastructure." In *Treaty Ports in Modern China: Law, Land, and Power*, edited by Robert Bickers and Isabella Jackson. Routledge, 2016.
Reisz, Emma. "Projecting the Road: Topological Photography on the Yunnan-Burma Frontier." *The Chinese Historical Review* 25, no. 2 (2018).
Relyea, Scott. "Settling Authority: Sichuanese Farmers in Early Twentieth-Century Eastern Tibet." In *Frontier Tibet. Patterns of Change in the Sino-Tibetan Borderlands*, edited by Stéphane Gros. Amsterdam University Press, 2019.
Rice, Stephen, André Roy, and Bruce Rhoads, eds. *River Confluences, Tributaries and the Fluvial Network*. John Wiley & Sons, 2008.
Rinaldo, Andrea, Marino Gatto, and Ignacio Rodriguez-Iturbe. *River Networks as Ecological Corridors: Species, Populations, Pathogens*. Cambridge University Press, 2020.
Roe, Alan, and Iftekhar Iqbal. "Riverine Environment." In *A Companion to Global Environmental History*, edited by John R. McNeill and Maulden. Wiley, 2025.
Ross, Corey. *Liquid Empire*. Princeton University Press, 2024.
Rowe, William T. "Economic Transition in the Nineteenth Century." In *The Cambridge Economic History of China*, edited by Debin Ma and Richard von Glahn. Cambridge University Press, 2022.
Sadan, Mandy. *Being and Becoming Kachin: Histories Beyond the State in the Borderworlds of Burma*. Oxford University Press, 2013.
Saha, Jonathan *Colonizing Animals: Interspecies Empire in Myanmar*. Cambridge University Press, 2022.

Saikia, Arupjyoti. "Imperialism, Geology and Petroleum: History of Oil in Colonial Assam." *Economic and Political Weekly* 46, no. 12 (2011).

Saikia, Arupjyoti. *The Unquiet River: A Biography of the Brahmaputra*. Oxford University Press, 2019.

Saikia, Yasmin. *Fragmented Memories: Struggling to Be Tai-Ahom in India*. Durham University Press, 2005.

Sankaran, Chitra. *Women, Subalterns and Ecologies in South and Southeast Asian Women's Fiction*. University of Georgia Press, 2021.

Sari, Betti Rosita. "Trade Activities in Thailand Border Area Case Study: Mae Sai and Chiang Khong." In *Borders and Beyond: Transnational Migration and Diaspora in Northern Thailand Border Areas with Myanmar and Laos*, edited by Betti Rosita Sari. Yayasan Pustaka Obor Indonesia, 2018.

Schendel, Willem van. *Bengal Borderland: Beyond Stage and Nation in South Asia*. Anthem, 2004.

Schendel, Willem van. "Geographies of Knowing, Geographies of Ignorance: Jumping Scale in Southeast Asia," *Environment and Planning D: Society and Space* 20, no. 6 (2002).

Schendel, Willem van. *A History of Bangladesh*. Cambridge University Press, 2020.

Schendel, Willem van. "Origin of Burma Rice Boom." *Journal of Contemporary Asia* 17, no .4 (1987).

Schendel, Willem van. "Southeast Asia: An Idea Whose Idea Is Past?" *Bijdragen tot de Taal-, Land- en Volkenkunde* 168, no. 4 (2012): 497–503.

Schrader, Heiko. "A Himalayan Trading Community in Southeast Asia." In *The Moral Economy of Trade: Ethnicity and Developing Markets*, edited by Hans-Dieter Evers and Heiko Schrader. Routledge, 1994.

Scott, James C. *The Art of Not Being Governed*. Yale University Press, 2010.

Scott, James C. *In Praise of Floods: The Untamed River and the Life It Brings*. Yale University Press, 2025.

Sen, Sudipta. *Ganges: The Many Pasts of an Indian River*. Yale University Press, 2019.

Sen, Tansen. *India, China and the World: A Connected History*. Rowman and Littlefield, 2017.

Seshan, Radhika. *Narratives, Routes and Intersections in Pre-Modern Asia*. Routledge, 2016.

Sharma, Jayeeta. *Empire's Garden: Assam and the Making of India*. Duke University Press, 2011.

Shiva, Vandana. *Water Wars*. Pluto, 2002.

Simms, Peter, and Sanda Simms. *The Kingdoms of Laos: Six Hundred Years of History*. Curzon, 1999.

Simpson, Thomas. "Find the River: Discovering the Tsangpo-Brahmaputra in the Age of Empire." *Modern Asian Studies* 58, no. 1 (2023).

Skinner, G. William. "Cities and the Hierarchy of Local Systems." In *The City in Late Imperial China*, edited by G. William Skinner. Stanford University Press, 1977.

Skinner, G. William. "Regional Urbanization in Nineteenth-Century China." In *The City in Late Imperial China*, edited by G. William Skinner. Stanford Univerity Press, 1977.

Song, Nianshen. *Making Borders in Modern East Asia: The Tumen River Demarcation 1881–1919*. Cambridge University Press, 2018.

Soonthornpasuch, Suthep. "Islamic Identity in Chiengmai City: A Historical and Structural Comparison of Two Communities." PhD diss., University of California, 1977.

Stargardt, Janice. "Hydraulic Works and South East Asian Polities." In *Southeast Asia in the 9th to 14th Centuries*, edited by D. Marr and A. Milner. ISEAS-Yusof Ishak Institute, 1986.

Stevens, Keith. "A Tale of Sour Grapes: Messrs. Little and Mesny and the First Steamship Through the Yangzi Gorges," *Journal of the Hong Kong Branch of the Royal Asiatic Society* 41 (2001).

Strang, Veronica. "Common Senses: Water, Sensory Experience and the Generation of Meaning." *Journal of Material Culture* 10, no. 1 (2005): 92–120.

Strang, Veronica. "Fluid Consistencies: Material Relationality in Human Engagement with Water," *Anthropological Dialogues* 21, no. 2 (2014): 133–50.

Stuart-Fox, Martin. *A History of Laos*. Cambridge University Press, 1997.

Tagliacozzo, Eric. "Ambiguous Commodities, Unstable Frontiers: The Case of Burma, Siam, and Imperial Britain, 1800–1900." *Comparative Studies in Society and History* 46, no. 2 (2004).

Tagliacozzo, Eric, and Wen-Chin Chang. "The Arc of Historical Commercial Relations Between China and Southeast Asia," in *Chinese Circulations: Capital, Commodities, and Networks in Southeast Asia*, edited by Eric Tagliacozzo and Wen-Chin Chang. Duke University Press, 2011.

Tagliacozzo, Eric, and Wen-Chin Chang, eds. *Chinese Circulations: Capital, Commodities, and Networks in Southeast Asia*. Duke University Press, 2011.

Tagliocozzo, Eric, Helen F. Siu, and Peter C. Perdue, eds. *Asia Inside Out: Connected Places*. Harvard University Press, 2015.

Taillant, Jorge Daniel. *Glaciers: The Politics of Ice*. Oxford University Press, 2015.

Taylor, Philip. "Water in the Shaping and Unmaking of Khmer Identity on the Vietnam-Cambodia Frontier." *TRaNS: Trans -Regional and -National Studies of Southeast Asia* 2, no. 1 (2014).

Taylor, Philip. "The Cosmopolitan Delta: Ethnic Pluralism at the Mouth of the Mekong." In *Routledge Handbook of Contemporary Vietnam*, edited by Jonathan D. London. Routledge, 2022.

Thao, Le Ba. *Vietnam: The Country and Its Geographical Regions*. Gioi, 2017.

Theobald, Ulrich. "Southwest China: Local Conditions and Economic Trajectories." In *Southwest China in a Regional and Global Perspective*, edited by Ulrich Theobald and Cao Jin . Brill, 2018.

Thitibordin, Amnuayvit. "Control and Prosperity: The Teak Business in Siam 1880s–1932." PhD diss., University of Hamburg, 2016.

Toyota, Mika. "Cross-Border Mobility and Social Networks: Akha Caravan Traders." In *Where China Meets Southeast Asia: Social and Cultural Change in the Border Regions*, edited by Grant Evans, Christopher Hutton, and Kuah Khun Eng. Institute of Southeast Asian Studies, 2000.

Tsing, Anna. *Friction: An Ethnography of Global Connection*. Princeton University Press, 2004.

Tun, Sai Aung, *History of the Shan State: From Its Origin to 1962*. Silkworm Books, 2009.

Tun, Than. ed. *Wekmasuk Wundauk U Latt's Diary [1888–1889]*. Vol. 2. Busan University of Foreign Studies Press, 2014.

Turner, Sarah. "Borderlands and Border Narratives: A Longitudinal Study of Challenges and Opportunities for Local Traders Shaped by the Sino-Vietnamese Border." *Journal of Global History* 5, no. 2 (2010).

Turner, Sarah. "Slow Forms of Infrastructural Violence: The Case of Vietnam's Mountainous Northern Borderlands." *Geoforum* 133 (July 2022): 189–97.

Turner, Sarah, and Jean Michaud. "'Weapons of the Weak': Selective Resistance and Agency Among the Hmong in Northern Vietnam." In *Agrarian Angst and Rural Resistance in Contemporary Southeast Asia*, edited by Dominique Caouette and Sarah Turner. Routledge, 2009.

Walker, Andrew. "Conclusion: Are the Mekong Frontiers Sites of Exception?" In *On the Borders of State Power: Frontiers in the Greater Mekong Sub-Region*, edited by Martin Gainsborough. Routledge, 2009.

Walker, Andrew. *The Legend of the Golden Boat: Regulation, Trade and Traders in the Borderlands of Laos, Thailand, China, and Burma*. Curzon, 1999.

Walker, Anthony R. "The Lahu People: An Introduction." In *Highlanders of Thailand*, edited by John McKinnon and Wanat Bhruksasri. Oxford University Press, 1986.

Wang, Yaxiang, Ziyi Cao, Zhaojun Pang, et al. "Influence of Three Gorges Dam on Earthquakes Based on GRACE Gravity Field." *Open Geosciences* 14, no. 1 (2022): 453–61. https://doi.org/10.1515/geo-2022-0350.

Webber, Michael E., *The Thirst for Power: Energy, Water, and Human Survival*. Yale University Press, 2016.

Wheeler, Charles. "One Region, Two Histories: Cham Precedents in the History of the Hội An Region." In *Viet Nam: Borderless Histories*, edited by Nhung Tuyet Tran and Anthony Reid. University of Wisconsin Press, 2006.

Whitcombe, Elizabeth. *Agrarian Conditions in Northern India*. University of California Press, 1972.

White, Richard. *The Organic Machine: The Remaking of the Columbia River*. Hill and Wang, 1996.

Whitington, Jerome. *Anthropogenic Rivers: The Production of Uncertainty in Lao Hydropower*. Cornell University Press, 2019.

Wilson, Jon. *India Conquered: Britain's Raj and the Chaos of Empire*. Simon & Schuster, 2016.

Winichakul, Thongchai. *Siam Mapped: A History of the Geo-Body of a Nation*. University of Hawaii Press, 1997.
Wittfogel, Karl. *Oriental Despotism: A Comparative Study of Total Power*. Yale University Press, 1957.
Wong, Soo Mun Theresa. "Making the Mekong: Nature, Region, Postcoloniality." PhD diss., Ohio State University, 2010.
Worster, Donald. *Rivers of Empire: Water, Aridity, and the Growth of the American West*. Oxford University Press, 1992.
Wouters, Jelle. "Keeping the Hill Tribes at Bay: A Critique from India's Northeast of James C. Scott's Paradigm of State Evasion." *European Bulletin of Himalayan Research* 39 (2012): 41– 65.
Wouters, Jelle. "Relatedness, Trans-species Knots and Yak Personhood in Bhutan Highlands." In *Environmental Humanities in the New Himalayas: Symbiotic Indigeneity, Commoning and Sustainability*, edited by Dan Smyer Yü and Erik de Maaker. Routledge, 2021.
Yang, Bin. *Between Winds and Clouds: The Making of Yunnan*. Columbia University Press, 2008.
Yang, S., K. H. Xu, J. D. Milliman, H. F. Yang, and C. S. Wu. "Decline of Yangtze River Water and Sediment Discharge: Impact from Natural and Anthropogenic Changes". *Scientific Reports* 5, article no. 12581 (2015). https://doi.org/10.1038/srep12581.
Yegar, Moshe. *The Muslims of Burma: A Study of a Minority Group*. Verlag Otto Harrassowitz, 1972.
Yü, Dan Smyer. "Symbiotic Indigeneity and Commoning in the Anthropogenic Himalayas." In *Environmental Humanities in the New Himalayas: Symbiotic Indigeneity, Commoning and Sustainability*, edited by Dan Smyer Yü and Erik de Maaker. Routledge, 2021.
Zhang, Ling. *The River, the Plain and the State: An Environmental Drama in Northern Song China*. Cambridge University Press, 2016.
Zhang, Peng, Yani Najman, Lianfu Mei, et al. "Palaeodrainage Evolution of the Large Rivers of East Asia, and Himalayan-Tibet tectonics." *Earth-Science Reviews* 192 (2019): 601–30.
Zhang, Qiang, Zexi Shen, Yado Pokhrel, et al. "Oceanic Climate Changes Threaten the Sustainability of Asia's Water Tower." *Nature* 615 (2023): 87–93. https://doi.org/10.1038/s41586-022-05643-8.
Zvelebil, K. V. *The Lord of the Meeting Rivers: Devotional Poems of Basavanna*. Motilal Banarsidass; UNESCO, 1984.

Internet sources
GRID-Arendal. "River Discharge of Freshwater into the Mediterranean." https://www.grida.no/resources/5897. Accessed February 6, 2025.
Mekong River Commission. "Mekong Basin: Geography." https://www.mrcmekong.org/about/mekong-basin/geography/. Accessed February 6, 2025.

Richardson, Michael. "Dams in China Turn the Mekong into a River of Discord." YaleGlobal Online. https://archive-yaleglobal.yale.edu/content/dams-china-turn-mekong-river-discord. Accessed January 25, 2025.

Salemink, Oscar. "The Regional Centrality of Vietnam's Central Highlands." *Oxford Research Encyclopedia of Asian History.* Oxford University Press, 2018. https://doi.org/10.1093/acrefore/9780190277727.013.113.

Tsing, Anna L., Jennifer Deger, Alder Keleman Saxena, and Feifei Zhou, eds. *Feral Atlas.* Stanford University Press, 2020. https://feralatlas.supdigital.org. Accessed February 6, 2025.

U.S. Geological Survey, "EarthView—Three Gorges Dam Brings Power, Concerns to Central China", November 17, 2016. https://www.usgs.gov/news/science-snippet/earthview-three-gorges-dam-brings-power-concerns-central-china. Accessed February 6, 2025.

Woo, Ryan. "It's So Hot in China, Melting Glaciers Risk Collapsing Dams." *Sunday Morning Herald*, July 23, 2022. https://www.smh.com.au/world/asia/it-s-so-hot-in-china-melting-glaciers-risk-collapsing-dams-20220723-p5b3y2.html. Accessed Febuary 6, 2025.

WWF-UK. "The Yangtze", https://www.wwf.org.uk/where-we-work/places/yangtze. Accessed February 6, 2025.

Xinjinzhe, NJU, "Two Years After Yangtze Ban, Its Fishers Are Still Reeling", *Sixth Tone*, November 27, 2021, https://www.sixthtone.com/news/1009065#:~:text=According%20to%20the%20Yangtze%20River,roughly%20halving%20in%20the%201990s. Accessed February 6, 2025.

INDEX

Page numbers in *italics* refer to figures.

Abor (Adi) people, 13, 24; agriculture, 31; inter-ethnic relations, 41; mobility, 10; relations with British administration, 31–34; relations with Chinese, 31; relations with Tibetans, 33, 37; riverine settlements of, 32; as trading partner, 33, 40, 55
actor-network theory. *See* Latour, Bruno
affectivity: commoning process, 102; human bonding with mule, 163; nature of mobility, 18
Africa: as part of Silk Route, 176; West African in Assam, 39
Akha people, 98–99; trade and intermarriage with Hui muleteers, 166
America and Americans: advance in China's east coast, 136–138; boat voyaging compared with Chinese, 144; cigarettes, 60; engineers, 48; hydrological projects compared to Chinese, 175; intelligence, 143; missionaries, 59; ships, 93; steamers, 137
Amu Darya River, 4
Andaman Sea, 4, 44, 70
Anglo-Burmese Wars, 11–12, 48, 70–71, 73

Annales School, 7–8
Anthropocene, 16, 173, 175, 180–181
Asia-Pacific (and Pacific) Ocean, 5, 16, 136, 152, 176
Assam: annexation of, 11; Assam-Bengal links, 39; Chengdu-Assam route, 132; Chinese interest in, 53; dams in, 178; and Ledo Road, 155; silk from, 35; tea plantation in, 24; as transregional node, 11, 21, 24–29, 40, 124, 126, 156

Bangkok: British interests in, 95; Chinese commerce in, 98; higher commodity prices in, 156; as intra-Asian crossroad, 81; and Mekong commercial network, 91–93, 100; Moulmein connections, 68–69; as trading node, 69; Yunnan connections, 152
Bassac River, 91
Batang, 24, 26, 78; as transregional node, 28, 34, 42, 131–132, 155, 210n
Bay of Bengal, 11, 17, 21, 42, 171
Beijing, 28, 35, 143
Belgian products, 78
Belt and Road Initiative, 176

245

Bengal, 4; Assam connections, 39; boats in, 51; as commercial hub, 24, 27; jute, 38, 58; railway line in, 27; Tibet connections, 25–26; as transregional node, 11, 20–21, 25, 29, 104

Bengali people, 38; in Thailand, 81

Bhamo: Bhamo-Dali trade route, 156, 162; Chinese influence in, 52; as confluence zone, 8, 100; contested Chinese and British interest in, 54; cultural plurality in, 59–60; as Hui presence in, 60; and Kachins, 167; Sirajganj compared with, 49; trade of, 48–49, 134; transregional node, 27, 57, 78, 83, 113, 114, 128, 133, 155; Yangzi connection, 132; water discharge at, 178

Black Flags, 13, 120–122, 140, 167

Black River, 89, 113–114, 122; French missionaries on, 116; local powers along, 121

boatmen: and mnemonic practice, 145; "as more beautiful than god," 214n; skills of, 107, 144, 145

boats: cheaper than steamers, 51; steamers compared with, 144; as tools of resistance, 75–76, 121

Brahmakunda, 40

Brahmaputra: Chinese interest in, 53; as cosmopolitan space, 41–42; decommoning of, 177-178; historiography of, 20–21; nomenclature of, 187n; trade routes with Yangzi valley, 37; tributaries of, 21–23

Braudel, Fernand, 5, 9, 168

British Empire, 21, 69; commercial limits of, 51; and engagements with borderland people, 30, 36, 74; transregional connections independent of, 42; vulnerability of, 86

Buddha: invocation of, 214n; millenarian, 99; statue, 59; tooth of, 82

Buddhism: missionaries, 11; monasteries and temples, 82, 146, 150; monks, 99

buffalo, 79, 101

Calcutta (Kolkata), 11, 30, 25, 40; and Canton, 27; Port, 20, 58; and Shanghai, 26, 28; Yangzi, 132; and Yibin, 132

Canton, 11; and Assam, 38; and Calcutta, 27; as transregional node, 72

capital, 31, 48, 72, 173, 179

caravans and caravanners, 19, 71, 89, 102, 114, ; and Bhamo, 48, 156; and human network, 165–169; Kachin protection of, 55; Karen relations with, 82; and Lu people, 101; and Manhao, 112; and Mekong River, 156; as metonym for river, 157; profitability of, 155, 156; and Red River, 155; resting places, 162; routes, 153; scaling heights, 160

Cartesian dualism, 173

Caspian Sea, 26

cat: as metaphor of China, 138; as simile for gorge, 146

Central Asia, 5, 26, 27

Cham Muslims, 99–100

Chao Phraya River, 13, 61, 66, 68, 71, 81, 82, 88, 91, 97–100, 109, 156

Chiang Mai, 68–69, 91, 98; as cosmopolitan space, 102–103

Chittagong, 11; Chittagonians in Assam, 39; Chittagonians in Rangoon, 58; Hill Tracts of, 104; revival of Port of, 129; ship crew from, 58

Chongqing, 9, 130–135

Christianity, 10; fear of conversion to, 29, 33, 65; toleration of, 59; Yunnan Muslim attitude towards, 123

climate change, 176, 180

commons and commoning, 3, 7, 14 –16, 42, 62, 87, 126, 149, 170

confluence (*sangama*): as actant and sites of mobility, 8; of Chao Phraya, 86; and commoning, 3; Ganga-Brahmaputra, 24; of Irrawaddy, 44, 46, 54, 178; and limits of ethnic autonomy, 171; lords of, 62; and market place, 83; Mekong, 91, 106; in

Sadiya, 24; of Salween, 60; of Red, 111; of Yangzi, 113, 129; and Zayat, 84
cosmopolitan and cosmopolitanism: of Brahmaputra, 38–42; ethnic, 40; of Irrawaddy, 58–62; of Mekong, 102–108; of Red, 122–126; of Salween, 80–86; of Yangzi, 165–168; Yunnan as, 9
Cotton Route, 99, 176
cow, 60, 69
Curzon, Nathaniel, 27–28, 172

Dali, 50, 89, 111, 113; as caravan hub, 29; route distances from river ports, 28; as transregional node, 66, 114, 133–134, 210n
dams: and loss of inclusive river space, 175; and sedimentation, 174–175; and tectonic faultlines, 174
decommonization, 15, 173–177
Delhi, 34
Deo clan, 121
Dhaka (Dacca), 4, 30, 42, 111, 152
Dibrugarh, 24, 39, 42, 78
dog, 35, 37, 86

East China Sea, 4
ecosystem services, 1, 176
elephant, 68, 69; conflict over, 75; mule compared with, 159; and teak transport, 75; war-time demand for, 157
El Niño, 177
Emerson, Ralph Waldo, 8, 181

Fluvial Asia, 5, 7, 174, 181
fluvial cosmos, 18
France: Haiphong-Kunming railway, 155; imperial activities in Indochina, 95–98; imperial advance and limits in Vietnam, 115–118; influence on the Yangzi, 137

Ganga River, 4, 20
Garo, 10

Germany, 78, 266; influence on the Yangzi, 117, 137, 150
glacial melting, 177
Goalundo, 23–24, 25
Gosain (trading pilgrims), 25
Gulf of Martaban, 65, 72, 74, 80, 87
Gulf of Tonkin, 109, 111, 114
Gurkhas, 25, 30
Guwahati, 24, 30, 39

Haiphong, 111–112, 117, 121, 155
Hanoi, 5, 12, 111–112, 117, 121, 156
Heidegger, Martin, 18, 107
Himalayas, 6, 8, 21, 28, 132, 219n
Hmong people, 100, 122, 166
Ho Chi Minh City, 4, 152
Holocene, 1
Hong Kong, 12, 40, 58, 93, 111, 112, 156
horse, 25, 134, 155; epidemics of, 32; cart driven by, 39; compared with mule, 161; in Batang, 210n
Hui (Panthay) Muslim: caravanners, 102–103, 157; as dominant muleteers, 49–54, 215n; mobility skills of, 158; rebellion, 49, 142–143; settlement patterns, 13, 52–54; speed, 158; trade and religious networks, 50–52
Humboldt, Alexander von, 148

Indian Ocean, 5, 16, 39, 50, 52, 83, 91, 136, 152, 156, 176
Indo-Pacific, 4, 6, 110, 170, 171
Indus River, 4, 26
inter-Asian trade, 27–30, 36–37
Irrawaddy: compared with other rivers, 43; cosmopolitanism of, 58–62; economic activities along, 48–49; formation of, 44; as national space, 43–44; nodal points of, 48–49; decommoning of, 178; nomenclature of, 192n
Irrawaddy Flotilla Company, 50–51; decline of, 51, 58
Islam, 140. *See also* Muslims

248 Index

Japan: commercial and strategic influence on Yangzi, 137–138; and Hui resistance, 142; impact of invasion on ethnic relations, 56; impact of invasion on Irrawaddy, 51; impact on mule traffic, 153–155; Japanese ladies in Bhamo, 59; and Mekong, 97; migrations due to invasion from, 81; silk from, 86; supplier of raw cotton, 113

Kachins, 13, 54–57; British, Chinese and Burmese appeasement of, 55–56; as managers of ferry crossings, 55; multiple identities of, 61; as owners of jade mines, 55
Karenni, 75–76
Karens, 13, 74–75
Khasis, 4; language of, 104–105
Khorat Plateau, 98, 99
Kolkata (Calcutta): 25–26, 40, 42; port, 20; and Shanghai, 28; as transregional node, 27
Konbaung dynasty, 49, 57, 95
Korea, 138
Kunlon, 66, 58, 153, 156
Kunming, 90, 96; and Assam, 155; and Batang, 210n; cosmopolitan, 165; decline in rainfall in, 177; French railways to, 118; as healthy place, 79; as transregional node, 109, 112, 114, 134, 155, 166; Washington DC compared with, 89

labor (animal), 169, 171
labor (human), 129, 145, 148, 169
Lao Cai, 111–112, 116, 120, 122, 155
Lao people, 13, 101–102, 108
Lamet people, 100–101
La Niña, 177
Lashi people, 143
Latour, Bruno, 8, 171
Ledo Road, 155
Leipzig, 131

Lhasa, as transregional node, 24, 25, 28, 31, 128, 131, 132, 155
Lijiang, 26, 28, 29, 38, 42, 60, 132, 143
Lisu people, 82–83
Lolo people, 119, 125, 143, 165, 167
longue durée, 5
Luang Prabang, 91; Chao Phraya connection, 66; as confluence zone, 8, 100, 102; as cosmopolitan space, 103; as trade center, 78; as transregional node, 98, 114, 166

macroregion, theory of, 135, 211n
malaria: immunity from, 56, 79, 119; and railways, 118
Mandalay: cosmopolitanism of, 58–59; Hui presence in, 166–167; Kachin political ambition over, 57; as transregional node, 25
Man Hao, 9, 111–113, 11, 119, 123
Marco Polo, 79, 95
Mediterranean Mountains, 9, 14
Mediterranean Ocean, 5–6, 157, 176
Meghalaya, 104, 178
Mekong River: as cosmopolitan space, 102–108; decommoning of, 179; European perception of, 88; nodal points of, 89–93; nomenclature of, 201n
Mengzi, 112–113, 116, 118, 123, 125, 155, 160
methodological regionalism, 18
Mishmis, 10, 13, 34–37; as bridge-maker, 37; centrality of trade among, 36; clans of, 35; commodities traded by, 35; ethnic groups engaged with, 40; resistance to the British by, 35
mnemonic practice, 145–148
Mon Khmers, 10
mosquitos, 119
Moulmein (Mawlamyine), 4, 65; cosmopolitanism of, 80–81
mule, 9–10, 48–49, 69, 112–113, 154–155; bonding with humans, 162–164; and Braudel, 16; breeding zones, 27; carrying

capacity of, 154–155; compared with other pack animals, 159–160; competition with railways, 157; complementary to rivers, 19, 118, 153; connecting India, China and Tibet, 132; population, 153; routes, 155–156; saddles, 163; skills and temperament, 159–163; traditional treatment of, 163; war-time demand of, 155

muleteers: commercial and cultural networks of, 165–167, 217n; settlements beyond Yunnan, 167

Muslims, 39, 60; in Assam, 39; in Chiang Mai, 81; Hui caravanners, 102, 165, 215n; in Mekong Delta, 99; Punjabi Muslims on the Yangzi, 150; in Sichuan, 142; spiritual figures, 38; in Yunnan, 120, 123

Myitkyina, 49; Chinese interest in, 53

Naga people, 10

Nanking (Nanjing), Treaty of, 12, 150

nature: beauty of, 148; and Cartesian Dualism, 173; dialectical interchanges with, 10; exploitation of, 15; Heideggerian approach to, 18, 107–108, 175; and human lifeworld, 181; and imperial projects, 72; as intuitive, 16; mobility in, 8; protection of, 33; and the universal, 186; veneration of nature, 106

Nepal, 11

Netherlands, 89

New York City, Simao compared with, 89

Nguyễn dynasty, 95, 115

opium, 56, 89, 99, 113, 131, 166

Organic River, 171

oxen, 9, 68, 69, 112, 167; compared with mule, 159; as metaphor, 181

Pacific Ocean. *See* Asia-Pacific (and Pacific) Ocean

Pacific railroads, 136

pack animal: Braudel's view of, 168; companion, 163; complementary to river transport, 9, 68–69, 152–153; human-animal bonding, 168–169; labor, 168–169, 171

Pearl River, 4, 141

Pegu, 11, 71, 80

Peru, 25

Philippines, 93

Phnom Penh, 91, 93, 100, 111

plague, 159

ponies, 9, 48, 60, 69, 104, 132, 153, 159

Pu'er tea, 69, 89

Qing empire, Yangzi River and, 138–140, 153

Quran, 150

railways, 27, 28, 98, 118, 136, 144

Rangoon (Yangon), 11; cosmopolitanism of, 58

Red River: commodities traded along, 113–114; as cosmopolitan and affective site, 122–126; decommoning of, 179; nodal points of, 111–113; nomenclature of, 204n; as site of Vietnamese nationalism, 110; transregional networks of, 110–111

Rennell, James, 20

Rima: as crossroad, 28, 34, 37; as trading center, 24

river: as agency and network, 3, 7–10, 171, 180; as commons, 3, 14–16; corruption of human memory of, 221n; deep ecology of, 1; dialogic power of, 10; in environmental and imperial history, 16–17; and ethnic mobility, 10–11; ethno-morphology of, 173; freedom of, 173; human and social dimensions of, 1; memory of, 8, 185n; and mobility, 8; as political and social space, 3, 10–14; source of energy, 172; spatial scales of, 2, 4–7

river diversion projects, 175

Russia, 27, 71; British diplomatic overture to, 136–137

250 Index

Sadiya, 26, 31; as confluence zone and trading crossroad, 8, 23, 24, 35
Saigon, 91–93
Saigon, Treaty of, 12
salinity, 176
Salween: as commons, 86–87; as cosmopolitan space, 84–86; decommoning of, 179; historiography of, 65–66; as imperial pathways to China, 72–73; nodal points, 66–67; nomenclature of, 197n; Salween River, caravan traffic and ethnic interactions, 30, 35–36, 52; unnavigability of, 65
Schiller, Friedrich, 148
Scott, James C., 55, 184n
sea level rise, 176
Shanghai, 4, 11, 128–129; US influence on, 135
Shans, 13; caravans, 69, 71, 76–79; fashions of, 77; immunity from malaria, 79; map-making skills of, 78
Silk Route, 176
Simao, 69, 89–90, 114
Sirajganj: as commercial hub, 24; as confluence zone, 8
South China Sea, 4, 93
Southeast Asian Massif, 21
steamers, 123, 144
Straits Settlements, 138
Swedish steel, 161
Switzerland, 89
Sylhet, 61, 104

Tagore, Rabindranath, 5, 8, 185n
Taiping Rebellion, 13, 140–142
Tengchong, 44, 54, 60
Tennessee Valley Authority, 173
Three Gorges Dam, 174
Tianjin, Treaty of, 12
Tibetan Plateau, 4, 16, 177

Tibetans, 33; riverine spirituality, 83; as traders, 83
traded commodities in Brahmaputra basin, 25

Victoria Falls, 28

Wa people, 79–80
Washington DC, Kunming compared with, 89
West Bengal, 177–178
Western civilization, 42
Wittfogel, Karl, 6–7
World War I, 73, 133, 137, 138, 157
World War II, 28, 36, 40, 157, 163, 167
Wuhan, 9, 129–130

Xishuangbanna (Sibsungpanna) 89, 103–104, 119

Yangon, 4
Yangzi, 26; decommoning of, 180; hydrology of, 127; Mussulmann (Muslim) Point on, 213n; nodal points of, 127–132; nomenclature of, 208n; transregional reaches, 127–128
Yao people, 13
Yellow River, 4, 7
Yen Bai, 111
Yichang, 128–132, 137–139, 143–144
Yunnan, 4, 20, 87; connectivity to coastal cities via rivers, 152; topography of, 152

Zayat (traditional Burmese travelers' lodge), 83–84; as affective and cosmopolitan space, 144–150; French influence on, 137; German influence on, 137; gorges, 146–48
Zomia, 5, 13, 184n, 217

The authorized representative in the EU for product safety and compliance is:
Mare Nostrum Group
B.V Doelen 72
4831 GR Breda
The Netherlands

www.ingramcontent.com/pod-product-compliance
Lightning Source LLC
Chambersburg PA
CBHW031804220426
43662CB00007B/526